AF823164

SOILS, STONES AND SYMBOLS

CULTURAL PERCEPTIONS OF THE MINERAL WORLD

SOILS, STONES AND SYMBOLS

CULTURAL PERCEPTIONS OF THE MINERAL WORLD

Edited by

Nicole Boivin & Mary Ann Owoc

Routledge
Taylor & Francis Group
LONDON AND NEW YORK

First published in 2004 by UCL Press

This edition published 2012 by Routledge
2 Park Square, Milton Park, Abingdon, Oxon OX14 4RN

Simultaneously published in the USA and Canada by Routledge
605 Third Avenue, New York, NY 10017

First issued in hardback 2016

Routledge is an imprint of the Taylor & Francis Group, an informa business

Copyright © 2004 by Taylor & Francis.

All rights reserved. No part of this publication may be reproduced, stored in a retrieval system, or transmitted, in any form or by any means, electronic, mechanical, photocopying, recording, scanning or otherwise, without the prior permission in writing of Cavendish Publishing Limited, or as expressly permitted by law, or under the terms agreed with the appropriate reprographics rights organisation. Enquiries concerning reproduction outside the scope of the above should be sent to the Rights Department, Cavendish Publishing Limited, at the address above.

Notice:
Product or corporate names may be trademarks or registered trademarks and are used only for identification and explanation without intent to infringe.

You must not circulate this book in any other binding or cover and you must impose the same condition on any acquirer.

British Library Cataloguing in Publication Data
Boivin, Nicole
Soils, stones and symbols: cultural perceptions of the mineral world
1 Minerals – history 2 Minerals – religious aspects 3 Minerals – social aspects
4 Stone – religious aspects 5 Soils – religious aspects 6 Symbolic anthropology
7 Archaeology 8 Folklore 9 Mineral studies, Prehistoric
I Title II Owoc, Mary Ann
306

Library of Congress Cataloguing in Publication Data
Data available

ISBN 13: 978-1-138-15193-2 (hbk)
ISBN 13: 978-1-84472-039-2 (pbk)

Cover image © Rick Bury

To Michael

To Sister Stephanie

PREFACE

Back at the beginning of the current decade, the two editors of this volume were working in isolation on separate continents, without anyone to talk to about their rather unorthodox interest in the symbolic and social aspects of soil and sediment use. We are therefore extremely grateful to Andy Jones, whose introduction precipitated a mutually stimulating interaction that eventually led to the production of this volume. Along the way there has been a conference session, held at the European Association of Archaeologists Meeting in Thessaloniki, Greece in 2001. That session was instrumental in demonstrating to us that, not only were we not alone in our interest in the non-utilitarian aspects of the mineral world, but in fact there was a great deal of enthusiasm amongst our colleagues for the topic. We are grateful to the participants in the original conference session entitled 'Soils, Stones and Symbols: Theorising the Mineral World', to John Chapman, who acted as a discussant, and to the many people who expressed an interest in the session, and joined in with the vigorous discussion it elicited. We are also grateful to all those who met our subsequent decision to turn the session into an edited volume with such enthusiasm, and agreed to contribute papers. These include a significant number of individuals who did not participate in the original session, and whose papers have been instrumental in enlarging the geographical and subject-matter scope of the volume. It has been a tremendous pleasure to both of us to discover that so much active and exciting research is taking place in a subject area that we didn't really think existed.

We would like to take this opportunity to thank all those who have helped see this project to fruition. We are grateful to our contributors for their hard work and patience, and to those who, for one reason or another, were unable to participate in the volume despite their enthusiasm for the project. We would also like to thank all those who provided valuable contacts, information, advice and encouragement during the production of the volume, including Richard Bradley, Mark Devitt, Jane Huber, Rosemary Joyce, Alan McPherron, David Pedler, Nick Saunders, Chris Scarre, and Paul Taçon. We are particularly grateful to Andy Jones and Michael Petraglia for going beyond the call of duty in the provision of ongoing support and advice. UCL Press has been extremely supportive of this project from the moment we contacted them, and we would like to acknowledge our great debt of gratitude to all those staff members who have helped in various ways to produce the volume, including Jeremy Stein and Sonny Leong. We would in particular like to thank Ruth Massey for her patience, perseverance, valuable advice and careful editing. We also thank all those who have granted permission to reproduce images appearing within the book.

Editors' Note:

Rather than ask our contributors to conform to a single dating system, we have instead chosen to have each chapter employ its author's (or authors') preferred notation (ie, BP, BC, BC (cal), etc).

Nicole Boivin and Mary Ann Owoc

November 2004

CONTENTS

NOTES ON CONTRIBUTORS

Nicole Boivin is a Research Fellow in Cognitive Archaeology at the McDonald Institute for Archaeological Research at Cambridge University. She has conducted research on both present-day and past societies in various parts of India, and is currently co-director of an international, multi-disciplinary research project aimed at understanding the origins of settled life in south India. Other research interests include archaeological theory, material culture, cognition, and ritual.

Adam Brumm is a doctoral candidate in archaeology at the Department of Archaeology and Natural History, Australian National University. He has conducted fieldwork in Australia, Indonesia and India, and is particularly interested in socio-symbolic aspects of prehistoric stone technologies. His doctoral research focuses on the lithic evidence for early hominin behaviour on the island of Flores in east Indonesia.

Jane E Buikstra, a biological anthropologist and archaeologist, is the Leslie Spier Distinguished Professor of Anthropology at the University of New Mexico. Her research interests include palaeopathology, human skeletal biology, palaedemography, forensic anthropology, genetic relationships within palaeopopulations, palaeodiet, and funerary archaeology.

Douglas K Charles is Professor of Anthropology and Professor and Director of Collections of Archaeology at Wesleyan University. His research interests include the archaeology of funerary practices and of complex gatherer-hunters, particularly in the American Midwest. He is the editor of *Theory, Method and Practice in Modern Archaeology* (with R Jeske, 2003, Praeger) and *Recreating Hopewell* (with J Buikstra, forthcoming, University Press of Florida).

Robert Frodeman is Chair of the Department of Philosophy and Religion Studies at the University of North Texas. He is the Director of the New Directions Initiative, editor of *Earth Matters: The Earth Sciences, Philosophy, and the Claims of Community* (2000, Prentice Hall), co-editor of *Rethinking Nature* (2004, Indiana), author of *Geo-Logic* (2003, SUNY), and Director of the Global Climate Change and Society Program.

Mary Ann Owoc is Associate Professor of Anthropology/Archaeology at Mercyhurst College. She has authored a number of articles on Neolithic/Bronze Age funerary monuments and prehistoric ritual. She does fieldwork on the Isles of Scilly, UK and in the eastern United States. Her research interests include the interrelationship between landscape and monument construction, Neolithic/Bronze Age perishable ceramic impressions, historic cemeteries, and the archaeology of communities.

Mike Parker Pearson is Reader in Archaeology at the University of Sheffield. His research interests include funerary archaeology, architecture and monumentality, the archaeology of ritual and religion, excavation methodology, ethnoarchaeology and public archaeology. He directs long-term archaeological projects in Madagascar and the Outer Hebrides. His ethnohistorical study of the materiality of monumental stone tombs in Madagascar has inspired his current field research, the Stonehenge Riverside Project.

David Robinson has participated in both CRM and academic archaeological research in the United Kingdom, Ireland, Spain, India, and California. His doctoral research at the University of Cambridge is focused upon the rock art of south-central California. Methodological and theoretical interests include the history of archaeology, indigenous perception, GIS and spatial analysis.

Nicholas J Saunders is Lecturer in Material Culture and British Academy Institutional Fellow at University College London. He has written many articles on Pre-Columbian symbolism and the anthropology of 20th-century conflict. He is the author of *Trench Art: Materialities and Memories of War* (2003, Berg), and editor of *Icons of Power: Feline Symbolism in the Americas* (1998, Routledge) and *Astronomies and Cultures* (with C Ruggles, 1993, University Press of Colorado).

Chris Scarre is Deputy Director of the McDonald Institute for Archaeological Research, Cambridge University, and Editor of the *Cambridge Archaeological Journal*. His research interests include the relationship of prehistoric monuments to their landscape settings and the use of colour by prehistoric societies.

Paul SC Taçon is a Principal Research Scientist in Anthropology at the Australian Museum. From 1995 to 2003 he was Head of the Australian Museum's People and Place Research Centre. He has conducted archaeological and ethnographic fieldwork since 1980 and has over 60 months' field experience in remote parts of Australia, Canada, southern Africa and the USA. Dr Taçon has co-edited three books and published over 95 scientific papers on prehistoric art, body art, material culture and contemporary Indigenous issues. He is a specialist in rock art, landscape archaeology and the relationship between art and identity.

Julieann Van Nest is a Senior Scientist (Geology, Geoarchaeology) at the New York State Museum. Her main focus is on advancing the routine integration of the earth sciences into archaeology, with special interests in North American mounds and earthworks, postglacial landscape evolution, and soil biomantles at archaeological sites.

LIST OF ILLUSTRATIONS

– 1 –

FROM VENERATION TO EXPLOITATION

HUMAN ENGAGEMENT WITH THE MINERAL WORLD

Nicole Boivin

At present writing, the power of the mineral world is difficult to ignore. Current global politics are dominated by conflicts that, regardless of stated claims, often have much to do with mineral resources, and in particular oil. Environmental disputes between nations and between corporations and individuals frequently revolve around the destructive effects of mineral exploitation activities and mineral by-products on ecological systems and human health. The age of global, media-perpetuated terrorism is dominated by fears of, amongst other atrocities, the release of radioactive or toxic minerals into urban ecosystems. And in an increasingly abstract and virtual capitalistic world, the price of gold nonetheless continues to serve as an important indicator of the health of the global economy, while diamonds remain important symbols of wealth in ever new social contexts. Minerals remain irrevocably linked to power, wealth and both local and global inequality.

The importance of the mineral world to human societies is nothing new, of course. The first archaeological evidence for mineral exploitation, in the form of Oldowan stone technologies, probably predates even the genus *Homo*, and observations of chimpanzee tool use would suggest that stone was one of the first materials used by evolving primates for making tools. Another key early technological breakthrough, the invention of pottery, is also related to the exploitation of a mineral resource – clay – that has subsequently been heavily used by human societies across the world for many millennia. Indeed, the centrality of mineral-based technologies to human developmental trajectories is acknowledged in the widespread division of prehistoric cultural sequences into stone, copper, bronze and iron-using phases. Minerals were also indispensable in the creation of the pyramids, palaces, tells, megaliths, earthen mounds, ancient mud brick cities and other monumental constructions of the ancient past, as well as the treasures they contain, that continue to amaze and inspire up to the present day.

However, while the importance of the mineral world may span all periods and regions of the globe, understandings of the mineral world clearly do not. While contemporary capitalist societies generally perceive minerals as commodities to be exploited for economic gain (Taçon, this volume), and, in a related fashion, as passive objects of disembodied scientific enquiry (Frodeman, this volume), pre-industrial societies often take a different view. Within such societies, minerals are frequently symbolically meaningful, ritually powerful, and deeply interwoven into not just economic and material, but also social, cosmological, mythical, spiritual and philosophical aspects of life. In fact, such societies do not even necessarily recognise 'mineral' as a distinctive category of matter, to be held in opposition to the so-called 'animal' and 'vegetal' realms of living matter.[1] Rather than neutral and inert, minerals in a number of societies are animate, and are engaged with on the basis of a worldview that understands all matter to be infused with spiritual energy and 'life-force' (to use a term employed by Taçon in this volume).

Nonetheless, aside from passing remarks and occasional tantalising studies, surprisingly little regard has been paid to non-functional aspects of the mineral world in either the anthropological or archaeological literature. While compendiums and detailed studies addressing social and symbolic aspects of human interactions with plants (eg Cloke and Jones 2002; Gosden and Hather 1999; Gremillion 1997; Rival 1998), animals (eg Gosden and Hather 1999; Ingold 1988; Manning and Serpell 1994; Wills 1990), landscapes (eg Ashmore and Knapp 1999; Bender 1993; Scarre 2002; Tilley 1994; Ucko and Layton 1999) and other elements of what Westerners typically define as the 'natural' world (eg Descola and Palsson 1996; Ellen and Fukui 1996; Ingold 1986; Strauss and Orlove 2003) can now be found in relative abundance, there has until now existed no equivalent volume relating to the mineral world.

This lacuna is all the more surprising in the context of the discipline of archaeology, which as Charles *et al* (this volume) point out, deals almost exclusively with the minerals found in an archaeological record that disposes of anything less durable. The oversight likely has much to do with theoretical paradigms in archaeology: while processual frameworks may have ignored the symbolic and ideological realms of human experience, post-processual studies have, at least until more recently, often stressed finished artefacts over technological processes and material substance. Focusing on the meanings inhering in symbolic artefacts and spaces, such studies have tended to overlook the materials from which they were created in the first place. Only more recently have some researchers begun to devote significant attention not only to artefacts themselves, but also to the flint, clay, quartz, chert, sand, obsidian, and chalk of which they are composed (eg Bender 1998; Jones 2002a, 2002b; Jones and MacGregor 2002; Owoc 2002; Parker Pearson and Ramilisonina 1998; Parker Pearson 2000, 2002; Taçon 1991; Thomas 1996, 1999).

This book is a response to the need to bring some of these still disparate studies together into a single volume in order to highlight the importance of new ways of addressing both the mineral and material worlds. It also represents an attempt to draw attention to the utility of breaking down traditional disciplinary distinctions that insist on the division of the archaeological record into separate and isolated spheres of study, thus precluding the holistic and integrated study of the substances and materials unearthed by archaeologists. Both the volume as a whole, and the individual studies contained within it, illustrate the complex ways in which the materiality of minerals links together monuments, rock art, technologies and landscapes into a unified story that contradicts traditional archaeological categorisations of the material world (for a related argument, see Parker Pearson and Ramilisonina 1998: 322). In addition, however, the volume also encourages the creation of new artefact categories, or at the very least draws attention to features of the archaeological record that are more usually overlooked by interpretive frameworks, such as visually unimpressive smudges of paint on stone (Robinson), unmodified pebbles and clasts (Saunders), surface textures (Scarre, Robinson), microscopic sedimentary layers (Boivin), and scatters of sand (Saunders) and sediments (Charles *et al*, Owoc).

Having described what the volume is about, it is important to note at the outset what the volume will not cover. While metals, water and certain biologically-derived materials like phytoliths and ash, and the more durable components of teeth and bones fall, strictly speaking, within the so-called category of 'mineral', the papers in this volume deal with these only incidentally, if at all. The decision to focus predominantly on stones and sediments was primarily practical, and had the aim of making manageable and coherent a theme that otherwise threatened to expand exponentially into ever wider subject areas and material culture realms. However, it was also felt that materials like water and metal present a number of unique characteristics that differentiate them from sediments and soils, and lead to unique sets of issues and symbolic practices (described, for example, in Bradley 2000; Childs and Killick 1993; Haaland *et al* 2002; Herbert 1993; Hosler 1994, 1995; Jones 2002a; Keates 2002; Levy 1999; Renne 1991; Richards 1996; Schmidt 1997; Saunders 2001b, 2002) that might more profitably form the focus of separate compendiums, at least at the present stage. This decision is supported by certain studies of classification in ethnographic and archaeological contexts. Postgate (1997), for example, has shown that in early Mesopotamian texts, metals were often treated separately from stone, though shells in contrast were identified as stones. He relates this to material properties and production patterns, including the fact that shell was widely used in third millennium Mesopotamia in the same ways as stone, while metal was not (*ibid*).

It should also be pointed out that, while it does contain ethnoarchaeological studies of material culture use in contemporary societies, and does reference a great deal of ethnographic and ethnohistorical literature, this volume does not include any studies by social or cultural anthropologists. While this omission relates in part to the disciplinary affiliation of the editors, it also reflects to some degree the lack of interest that contemporary anthropology has shown towards material culture. Recent studies of material culture by anthropologists are few and far between, with studies of art constituting a notable exception (eg Coote and Shelton 1992; Gell 1998; Küchler 1985, 1992; Morphy 1991; Munn 1986; Watson 2003).[2]

Despite these restrictions, the mineral world theme that is addressed in this volume by archaeologists from a range of backgrounds (as well as one philosopher of science) nonetheless encompasses a diverse array of issues that are arguably of interest to both archaeologists and anthropologists, as well as, potentially, scholars and professionals in a range of other fields, including the earth sciences, the philosophy of science, geography, and environmental studies. The remainder of this introductory chapter will explore some of these issues, drawing upon both the chapters in the volume and the wider archaeological and anthropological literature. More epistemological aspects of the study of the mineral world are addressed in Owoc's epilogue, as well as in the chapter by Frodeman.

THE ANIMATE EARTH

Perhaps the most consistent and salient point that emerges from a review of the papers in the volume and the wider archaeological and anthropological literature relating to the mineral world is that, for many people around the world, minerals are alive. While, as indicated above, and emphasised in particular in Brumm's chapter, most Westerners may perceive the mineral world as neutral, inert, and devoid of life, people in many other societies do not necessarily recognise such a firm distinction between mineral and non-mineral, animate and inanimate. This blurring of boundaries means that minerals in many societies are attributed with qualities and properties that most people in Western societies accord only to humans, animals, plants and/or the divine. In many contexts, this means that minerals in general, or in some cases certain minerals in particular, are recognised as animate, divine, powerful and/or sacred.

A number of anthropological and archaeological studies stress the animate qualities of stone, for example. Brumm's chapter highlights examples of stones that grow, get pregnant, and have souls and a gender in Australia and New Zealand. In Melanesian societies, ethnographers have recorded examples of stones believed to walk around, dance, light fires, transmit and cure disease, speak, procreate and kill (Kahn 1990; Roe and Taki 1999). In many societies across the world, as amongst the Chumash people of coastal California described in this volume by Robinson, certain stones are believed to contain mythic or other beings who were turned into stone, and who in some cases can on occasion change back into their human form (eg Brady and Prufer 1999; Hedges 1993; Kahn 1990; Roe and Taki 1999; Taçon and Ouzman 2004; Taçon 1991).

In his chapter, Saunders notes that this belief in the possibility that culture heroes can shift between human and stone forms is key to understanding why mountains, rock outcrops, and human modified stone and stone constructions have often held such significance for Andean peoples, in the past and right up to the present day. This type of conceptualisation is also central to Australian Aboriginal perspectives on the landscape, which is believed to embody the actions of Dreamtime Ancestral Beings. For Aboriginal peoples, notable geological features of the landscape are often understood as the final resting places of such Beings, where they turned into stone (Taçon 1991; see also Brumm, this volume). Accordingly, the modification of stone by humans can sometimes be seen not so much as a *creative* activity, but instead as a *release* of entities already residing within the stone (Jones and White 1988; Taçon and Ouzman 2004).

The animate qualities with which clay and earth are attributed are often similar, though not identical, to those just described for stone. While many traditions cross-culturally hold that clay is the material from which the first humans were made (see Boivin, this volume), it is the divine rather than human qualities of earth that seem more often to be emphasised (eg Barley 1994; Gosselain 1999). Stones may contain the spirits of ancestors or heroic individuals, but earth is the embodiment of a divine, all-encompassing creative force. Thus, for example, Brady and Prufer describe how for many Mesoamerican and Amerindian groups, the earth has traditionally been understood as a sacred and animate entity (Brady and Prufer 1999). For the Q'eqchi, the Earth God is the central deity amongst many, and is associated with rain, fertility and abundance (*ibid*). Parker Pearson and Ramilisonina (1998: 311) similarly note that for the Merina of Madagascar, the earth was considered so sacred that even the king could not build his house from it. Saunders observes in his chapter that the conceptualisation of the earth as a sacred, female, living entity is widespread across the whole of the Americas.

This personification of earth as female is another common pattern, and contrasts with the male spirits that are frequently understood to dwell within stone (see also Taçon 1990, 1991: 204). The notion of an Earth Goddess or Earth Mother is found in such diverse parts of the world as Africa (Barley 1994; Gosselain 1999; Lovell 1998), India (Boivin 2001; Jain 1984; Kinsley 1986), Australia (Taçon 1991; Watson 2003) and the Americas (Saunders, this volume; Sillar 1996). Such a conceptualisation likely derives from the fertile qualities of soil, and the fact that, as the Aboriginal narrator in Taçon's chapter emphasises, all things ultimately derive from the earth (see also Lovell 1998 and Watson 2003). According to the Earth Diver myth discussed in this volume by Charles *et al*, and common to a range of cultures in Europe, Asia and North America, the world itself only comes into being when an animal dives into the sea and finally successfully brings some soil to the surface, which expands to create the land.

The sacredness that is attributed to the creative powers of earth (which is sometimes just soil/clay, but at other times the entire lithosphere) explains why transgressions of it often demand special behaviour, rituals, and offerings (Burton 1984; Jain 1984). Robinson's chapter, for example, describes how inappropriate behaviour amongst native California Chumash while mining minerals from the earth could have disastrous consequences. In some cases, obtaining the right to take from the earth even demands sacrifices (eg Barley

1994). Sillar (1996), for example, notes that human sacrifices in the form of children were annually made by at least one Andean community of potters to appease the mountain deities who controlled their clay mines. The children were sealed alive within deep-shaft tombs (*ibid*). Saunders, in his chapter, describes how gold and silver mining in Inka times were accompanied by rituals that involved petitioning the mountain to yield its mineral wealth. Mining into mountains perpetuated a 'ritual encounter' that linked together the upper and lower worlds of Inka cosmology.

The use of tunnels into the ground to interact with Earth deities or spirits is also a common pattern across many cultures. For the Maya, for example, caves are considered points of access to the dwelling places of important deities and ancestral spirits (Brady and Prufer 1999). Accordingly, items originating in caves, like speleothems, are considered powerful, and are used in rituals and ceremonies (Brady *et al* 1997). For the indigenous people of the Highlands of Irian Jaya, caves, crevices and holes in the ground are considered sacred because they are the places where life emerges (Hampton 1999). Across Mesoamerica, caves are similarly closely associated with rain and fertility, both crucial life-giving components of successful agricultural practice (Brady *et al* 1997). However, caves and other links to the earth often have ambiguous connotations. For the Maya, caves can cure, but they can also bring sickness (Brady and Prufer 1999), and in Ireland, digging or disturbing the ground was avoided on certain days of the year because of the dangers associated with tampering with the ground when links to the spirit world were heightened (Cooney 1998).

MEANINGFUL MINERALS

With all of these links to life, spirits and the sacred, it is not surprising that minerals have in many contexts emerged as incredibly powerful cultural symbols. In a vast range of cultural contexts across the world, societies have drawn on the evocative qualities of minerals to signify a whole plethora of meanings and concepts. Given the mineral bias of the archaeological record, it is not surprising that the symbolic qualities of minerals have increasingly begun to attract the attention of archaeologists, who have slowly begun to recognise the implications of such associations to the patterning they observe in the archaeological record. One of the key pioneers in drawing archaeological attention to the significatory aspects of the mineral world has been Paul Taçon, whose discussion of the 'power of stone' in archaeological and Aboriginal societies in Australia has been enormously influential in archaeological circles (Taçon 1991; see also 1990, and Taçon and Ouzman 2004). Indeed, Australia has proven to be a rich source of anthropological and archaeological insights into the metaphorical properties of the mineral world for a significant number of researchers (eg Jones and White 1988; McBryde 1997a, 1997b, 2000; Tilley 1994; Watson 2003; also Brumm, this volume), though as the range of geographical regions covered in this volume indicates, the patterns observed there are far from unique.

Interestingly, and in further contradiction to the Western scientific understanding of minerals as inorganic and inert, many of the symbolic associations of minerals draw in one way or another on the body. For example, mineral sources and quarries are often metaphorically linked to body parts or emissions. In this volume, Brumm observes that for Aboriginal Australians, raw stones and minerals are in many cases thought to be the blood, bones, fat, pus, or internal organs of Ancestral Beings deposited onto the landscape and metamorphosed into rock. Saunders meanwhile notes in his chapter that the Desana of the Amazon regarded rock crystal as concentrated semen. Similar beliefs also characterise clay and ochre sources in many parts of the world. For example, the pit from which a Yoruba potter in Africa obtains clay is known as the womb or vagina of the goddess Iya Mapo (Barley 1994). For the Batammaliba of Togo and Benin Republic in Africa, the clay that is used to make mud houses is compared to flesh, while the clay-based plaster that is applied to its surface is referred to as 'skin' (Blier 1987: 119).

Links between clay and flesh are in fact common in the anthropological literature (eg Taçon 1990), as are associations between stone and bone. Robinson notes in this volume that one Native American tribe in California is recorded as referring to a white mineral used in healing and body painting as powdered human longbone. The mineral is obtained from a hole where the rock giant Yayali is said to have thrown the bones of people he devoured. Taçon (1990, 1991) has extensively discussed the links drawn between bone and stone amongst Australian Aboriginal peoples (see also McBryde 1997b), who often view stone sources and geological outcrops as the petrified remains of the bones of certain Ancestors. The bones of deceased humans were accordingly placed by some Aboriginal groups into clefts and holes in rocky outcrops, thus reinforcing the association between humans and Ancestors (Taçon 1991). Taçon (1990, 1991) notes that such groups view bone and stone as symbols of persistence and immortality because of their enduring quality. An association between bone, stone and the eternal also characterises traditional belief systems on the island of Madagascar (Bloch 1971; Huntington and Metcalfe 1979; Parker Pearson and Ramilisonina 1998). Human life in Madagascar is characterised as a process of 'hardening' (Bloch 1995a, 1995b), and death and its associated rituals transform soft, moist flesh into hard, dry bone, and ephemeral wood houses into the permanent stone tombs of the ancestors (Bloch 1971; Huntington and Metcalfe 1979).

Water, wood and other natural substances are not uncommonly drawn into symbolic relationships of contrast and similarity with soil and stone, creating the potential for an ever richer and more complex web of associations and meanings, as attested in Parker Pearson's chapter in this volume. Within such networks, the bringing together of contrasting substances is sometimes perceived as a powerful and even dangerous occurrence. In Australia, female earth and male rain together bring fertility and life (Watson 2003), an association that is echoed in South Africa, where Ndebele potters argue that female clay and male water have to sleep together for a night before ceramic production can begin (Gosselain

1999: 212). Taçon (1990, 1991) has argued that the union of stone and earth is a powerful symbolic metaphor in several parts of the world, including England and Australia. He sees the Cerne Abbas giant that is carved into chalk in Dorset, England and believed to possess the ability to help women conceive, as a portrayal of the power that derives from the unification of male-phallic-stone fertility with female-uterine-earth creativity (Taçon 1990). Robinson's discussion (this volume) of 'extended interfaces' in Chumash rock art, and the additional symbolic associations that are brought to bear on human engagement with powerful stones when mineral pigments are applied, implies a similar intensification of the power contained in mineral substances when they are combined. The beliefs described by Robinson for the Chumash of California invite comparison with those of the indigenous peoples of Australia, who believe that the application of bright ochre pigment to rock surfaces unites male and female symbols to produce an especially powerful object or image that is believed to resonate Ancestral power (Taçon 1991: 204–05). Context is important, though: the success of male-dominated activities like hunting requires that tools such as stone spear-points not be contaminated with female-linked substances like ochre (*ibid*).

Clearly, the gendered associations of mineral substances are, as the foregoing discussion suggests, an important aspect of the symbolic potency of the mineral world in a range of cultures. Aboriginal peoples in the Central and Western Desert of Australia ascribe gender to all landform elements, and identify the zone beneath the surface of the ground with women. Accordingly, ochres lying underground are called 'proper mother ones' while those on the surface of the ground are called 'boy ones' (Watson 2003: 75). The 'mother ochres' stay on the body and are therefore superior to 'boy ochres' as substances for ceremonial use (*ibid*). The relationship between gender and use is a common one, and often relates to the gendered roles pursued in mineral technologies involving the production of ceramics, mud-built architecture, stone tools, metal items and rock art, for example (Barley 1994; Blier 1987; Boivin 2001; Brumm, this volume; Burton 1984; Gosselain 1999; Hampton 1999; Huyler 1995; Jain 1995; Jones and White 1988; Lovell 1998; Sillar 1996; Taçon 1991). Because the good ochres in the Central and Western Desert are female, it is women who collect and prepare red ochre, giving some of what they have prepared to their husbands for use in ceremony. In sub-Saharan Africa, in contrast, it is only men who may 'break' the female earth, while women form it (Barley 1994). Accordingly, potting is a female craft amongst the Konso of Ethiopia, who link fertility, earth, and pots, the latter being associated symbolically with wombs The ambiguity such symbols can take on is indicated by the fact that potting is nonetheless a 'despised craft', regarded by the Konso as an 'unnatural' treatment of the earth (*ibid*: 53). Such patterns of gendered mineral engagement are the inspiration behind my own suggestion that the pink granite used for making grinding implements in Neolithic south India was associated with women, while the black dolerite that penetrates it in such an evocative way in the landscape, and is used for making axes and ithyphallic petroglyphs, was associated with men (Boivin 2004).

Mineral symbolism is used to define and negotiate not just gender roles, however, but also many other varieties of social distinction. Madagascar beliefs compare processes of biological and social aging to a gradual 'hardening', with the result that harder substances, like stone and very hard woods, are also linked with the stability and wisdom of older members of society (Bloch 1995a, 1995b). Taçon (1991: 194) meanwhile notes that stone axes in Australia were a symbol of not only masculinity but also age, and belonged to 'older' men, whose role as lenders of such tools to women and young men reinforced their status, position and power within the community. Minerals can also signify groups that cross-cut age and gender. For example, Kogi individuals in northern Columbia are associated through their lineage with certain minerals and colours, and the lineage itself is regarded as the owner of these minerals and their symbolic attributes (Saunders, this volume). Amongst the Nuba of south-eastern Sudan, the yellow and red ochres that are used to decorate the body are also associated with different patriclans (Faris 1972). Each patriclan has its own characteristic colour, and although the colours are all red or yellow, each clan has a place from which to gather its ochre, resulting in 'slight differences between yellows and reds which are ostensibly visible to members of the community' (*ibid*: 32).

As this and some of the other examples discussed here have shown (see also Cooney 2002; Jones 2002a; Owoc 2002; Petraglia *et al* 2002; Saunders 2002; Spence 1999), the symbolic associations of minerals are sometimes closely linked to colour. It is argued, for example, that some of quartz's symbolic potency derives from its white colour, which in the Western Christianised world embodies notions of peace, innocence, purity, virginity, goodness and life (Darvill 2002: 74). Saunders notes in his chapter that in the north-eastern United States, the Huron-Wyandot 'White Lion' mythic being carried white stones – including quartz, white chert, flint and chalcedony – that helped to assure success in hunting and courtship. Most other authors in the present volume also implicate colour in mineral symbolism in one way or another, including in particular myself, Charles *et al*, Owoc, Scarre, and Taçon. In their chapter, for example, Charles and his co-authors note that mineral symbolism may have been accentuated (along the lines of Robinson's 'multiple interfaces') through links with colours and the cardinal directions. Owoc meanwhile discusses the relationship between coloured clays and solstial movement and alignment. Scarre's chapter wisely cautions against interpreting archaeological patterns as colour-related, however, just because colour may be most obvious to us. Colour, as he points out, can also be a minor feature, only incidentally related to properties like texture and origin that may have been more salient to people in the past (regarding texture, see also Cummings 2002; MacGregor 2002; Owoc, this volume).

MINERAL BIOGRAPHIES

Scarre's point about the potential importance of mineral *origins* is an important one, and provides an opportunity to turn this discussion towards an analysis of the *technologies* associated with human engagement with the mineral world. While, as noted at the outset of this chapter, non-utilitarian aspects of minerals have long been neglected, it can be argued that recent years have seen the symbolic properties of the mineral world receive increased attention. This interest has come most notably from archaeologists, and is in part the result of a number of parallel developments in archaeology. In particular, it probably has much to do with the shift away from the exclusive interest in finished form that characterised many early symbolic studies of material culture (critiqued by, amongst others, Dietler and Herbich 1998, and Lemonnier 1992), a move that has led to avid exploration of both the social aspects of technology (Dobres and Hoffman 1999; Lemonnier 1992; Pfaffenberger 1988, 1992) and the biographies of things, places and monuments (Bailey 1990; Chapman 1994; Gosden and Marshall 1999; Jones 2002a, 2002b; Kopytoff 1986; Skeates 1995). Linked to this has been an increased appreciation of the evocative power of landscapes, and their implications for not just economic but also symbolic and ritual practice, particularly with respect to the creation and use of monuments and rock art sites (Bender 1998; Bradley 1998, 2000; Chippindale and Nash 2004; Tilley 1994, 1996). These developments have focused some degree of attention on minerals themselves, as the substances out of which the majority of preserved artefacts and landscape features are composed. As Charles *et al* note in their chapter, the mineral substances we often find deposited in monuments, graves and other contexts are probably in many cases at the final stage in a potentially long series of uses and transformations. This section aims to explore briefly some of the aspects of these mineral 'biographies'.

Taking from the earth

In order to be used by humans, minerals must first be obtained. Sometimes this involves little more than reaching out for a handy stone or strolling over to a nearby stream bank and scooping some clay into a basket. At other times, however, mineral acquisition involves a great deal of effort. Sometimes this effort is necessary because required mineral resources are far away, or deep within the earth, but just as often, it would seem, there is no good functional reason why mineral acquisition need be so strenuous and dangerous as it often is.

The common role of 'ritual journeys' in mineral acquisition provides a good illustration of this fact. In a wide variety of cultures, the value of a mineral is very often related as much to the journey that was made to acquire it as to the mineral itself. The importance of journeying is described in the chapter by Charles *et al*, who note that while the acquisition of rare or exotic minerals might have been one motive for travelling to far away places during the Middle

Woodland Period in North America, it was unlikely to have been the only one. Just as important would have been the prestige, power and esoteric knowledge obtained by journeying, and minerals may indeed only have been acquired to serve as proof that such a journey had actually taken place. The important role of journeys to mineral quarry sites in Australia has been extensively discussed by McBryde, who emphasises that Aboriginal groups travelled not only to acquire valued minerals like ochre, and stone for axes and grindstones, but also to fulfil ceremonial obligations and promote the education of young initiates (McBryde 2000, see also 1997a, 1997b). Journeys to mineral sources were often long and dangerous, and followed strictly prescribed routes (McBryde 1997a, 2000). In this volume, Robinson notes that for the Wintu tribes of western North America, the journey to an obsidian source was a 'semireligious quest' that entailed fasting by participants.

The spiritual and symbolic value of journeys made to acquire minerals often had much to do with the power of the source itself. Ethnographic and ethnohistorical writings, as well as in some cases ancient texts, indicate that in various regions of the world, certain mineral sources are believed to be extremely powerful places, whose potency carries into the substances taken from them and the uses subsequently made of them. Taçon, for example, records that according to Australian Aboriginal peoples, a stone tool's killing power comes from the stone source, and it is this power which saps the life out of its target (Taçon 1991: 203; see also Jones and White 1988). Aboriginal people would thus travel or trade over long distances to obtain symbolically powerful stone or ochre that was functionally equivalent to locally-available materials (Jones and White 1988; McBryde 1997b). Such a concern with mineral origins is not confined to Australia. Textual sources indicate that for ancient Mesopotamians, the *origin* of stone used in constructing palaces and temples was considered as important as more utilitarian qualities (Postgate 1997: 212). Similarly, Hampton (1999: 225) notes that the Dani of Irian Jaya distinguish stone blade types on the basis of not only structure and size, but also specific quarry source outcrops.

The importance of mineral sources themselves, and the power that is often attributed to them, is clear from various ethnographic and ethnohistorical sources that describe the prescriptions and proscriptions associated with quarry sites. Even in tribal and hunter-gatherer societies, quarries appear commonly to be owned by specific individuals, who do not, however, benefit economically in any significant way from this ownership. Instead, such positions are associated with social prestige and, particularly, social responsibility. In Australia, Aboriginal quarry custodians feared disgrace and even death if they did not fulfil their responsibilities to safeguard sacred quarry sites (McBryde 2000). Their permission was required before quarries could even be approached, let alone quarrying begun (Jones and White 1988; McBryde 2000). In both Australia and Papua New Guinea, ethnographers record that prescribed ritual activities had to take place prior to the initiation of quarrying, and women were frequently barred from approaching quarry areas (Hampton 1999; McBryde 1997b, 2000; McBryde and Watchman 1976; Burton 1984). Tungei informants attributed their success in obtaining stone from axe quarries to ritual purity and the correct axe-making magic (Burton 1984). Like initiation ceremonies, quarry expeditions in the Papua New Guinea highlands traditionally ended in 'ritual fights' in which women attacked the quarry camps, beating with sticks whatever lay in their paths, including the menfolk (*ibid*).

While archaeological investigations are seldom able to provide this level of ethnographic detail, archaeological studies are nonetheless increasingly providing tantalising suggestions that mineral sources were important to people in the past. Such evidence has come, in particular, from studies of stone used in the production of both axes and monuments in Western Europe. For example, physical studies of Neolithic axe sources in Britain have indicated that accessible sources were often overlooked in favour of more remote locations that were difficult and dangerous to reach, but may have been selected in part because they commanded particularly stunning views across the landscape (Bradley 1993, 2000; Bradley *et al* 1992; Bradley and Edmonds 1993). Bradley (2000: 86) observes that '[t]aken together, [axe] production sites are among the most remote archaeological monuments anywhere in England', and argues that such locations reflect an attempt to isolate axe production from the sphere of everyday activity during the Neolithic (*ibid*: 87). Bradley *et al* (2002: 232) argue that while some of the most suitable stone available was hardly used, axes made of poorer stone were carried across the Irish Sea, indicating that source location may have been more important than stone strength. Patton (1993) has similarly argued that the dolerite source used for making stone axes at Le Pinacle, located at the foot of a massive outcrop of granite, was likely a significant feature of the sacred geography of the Channel Islands, and selected as a source for this reason. Cooney (1998) has suggested that islands may have had some sort of special or sacred status that favoured their use as sources for stone axes despite their inaccessibility during the Neolithic period in Western Europe.

Studies of stone monuments have revealed similar patterns. In this volume, Scarre suggests that the carefully chosen stones in Neolithic monuments in north-western France may have referenced particular places in the landscape more than symbolic associations of colour or texture. Noting recent research by Bradley (2000), he discusses the well-known example of the bluestones at Stonehenge, known to have been transported from the Preseli Hills 240 km away in south-west Wales, as another potential example of this kind of phenomenon. Bukach (2003) has similarly argued that non-local stones were often employed in the construction of Middle Neolithic passage graves on Guernsey and Jersey, and may have referred symbolically to significant places in the landscape and the identities of groups associated with them. These and other recent investigations into the sources of stones used in creating megalithic monuments (see Darvill 2002; MacGregor 2002) are complemented by various studies suggesting that mineral sources may have been just as important in earthen monuments (Bradley 2000). In this volume, Charles *et al*, Parker Pearson and Owoc all draw attention to the earthen components of burial monuments, in both prehistoric America and Britain. Charles *et al* argue (see also Van Nest *et al*

2001) that the sod used to create the Hopewell monuments of the lower Illinois River valley was not the naturally available local sod, and may have been specially grown for use in the creation of monuments. Parker Pearson's chapter summarises a variety of examples of enigmatic, specially transported soils found in both monumental and settlement contexts throughout prehistory in England and Scotland.

Various other artefacts and items have formed the basis of similar conclusions about the significance of mineral sources to people in the past. Jones, for example, in addressing a range of artefact types in the British Isles, has argued for the importance of non-functional aspects of mineral sources, from the gold, jet, amber and other minerals used in Early Bronze Age jewellery (2002a) to the tempers obtained for incorporation into Neolithic ceramics on the island of Orkney (2002b). In the latter case, he argues that mineral tempers were obtained from significant places in the landscape that served to articulate relationships to ancestors and other communities (*ibid*). Recent research aimed at sourcing mineral pigments used in rock art has also sometimes revealed interesting patterns. It was found, for example, that in Lesotho and South Africa, 'classic' Bushfolk rock-paintings were made with exotic ochre that was obtained some 20 to 100 km away, and would likely have involved a considerable amount of social ingenuity and manipulation to obtain it (Taçon and Ouzman 2004: 51).

Making with minerals

Minerals, once obtained from their sources, subsequently become the focus of a range of technological procedures that often continue to account for the symbolic meanings they possess. Technological modifications of minerals, such as stone-working and ceramic production, are often ritualised or symbolically-laden processes (eg Jones and White 1988; Gosselain 1999; Sillar 1996) that draw upon the metaphorical qualities of minerals themselves. Jones' (2000, 2002b) petrographic analysis of Neolithic pottery tempers on Orkney, for example, revealed that rock sources from different locales were being deliberately combined in the production of ceramics, a procedure that Jones argues was implicated in both the expression of social identity and the symbolic reinforcement of links between different communities. Other archaeologists have similarly commented on the symbolic associations that may have accompanied the inclusion of particular mineral tempers in prehistoric pottery in Britain (including Parker Pearson in this volume). Sillar (1996: 276), for example, notes that there must have been something particularly meaningful about flint that led Neolithic and Bronze Age potters to go to all the trouble of using it as a temper, as this would have involved complex and time-consuming procedures that are far from efficient. Parker Pearson, in previous works, has drawn upon a wider argument about the association of stone with ancestors in prehistoric England to argue that stone-tempered pottery was produced for interactions specifically with ancestors at various types of ceremonial sites (Parker Pearson 2000, 2002; Parker Pearson and Ramilisonina 1998).

The notion that ceramic tempers might have been used by prehistoric peoples for such non-practical reasons is reinforced by ethnographic studies of ceramic production like those of Gosselain and Livingstone-Smith, which indicate that the selection of tempers by potters in various African communities is 'not determined by environmental, technical or functional requirements' (Livingstone-Smith 2000: 38; see also Gosselain 1998). In this volume, my chapter similarly demonstrates that soils used by village women to make the plasters that coat mud-built houses in rural Rajasthan in India are selected more for their symbolic associations than their functional efficiency. The purity of a soil is an important consideration in house construction and maintenance technologies in rural India (Boivin 2001 and this volume; Ayyar 1998), just as it appears to have been in the technologies associated with figurine production in ancient Mesopotamia (Postgate 1997) and temple construction in India in the ancient past (Kennet and Rao forthcoming). In rural India today, women who are pregnant or menstruating and hence impure are thus theoretically excluded from involvement in soil technologies like house maintenance (Boivin 2001).

'Technologies' of monument and rock art creation appear to involve similar symbolic and social concerns. Stones employed in the construction of megalithic monuments often appear to have been significant in their own right (Cooney 1998; Cummings 2002; Darvill 2002; MacGregor 2002), and various studies, including that by Scarre in this volume, have suggested that they were strategically deployed during monument construction in order to articulate particular relationships, both social and symbolic (Bradley 2000; Bukach 2003). In the lower Illinois River valley, the creation of Hopewell burial mounds seems to have involved minerals in a very particular set of temporal and spatial relationships, and to have been understood as a technological re-enactment of the initial creation of the cosmos, as well as the representation of extant social hierarchies (Charles *et al*, this volume). The ritualisation of technological procedures is also found in the practices associated with the transformation of mineral pigments into paint by San (Bushmen) in South Africa. This transformation was accompanied by various ritual procedures, and necessitated, for example, the heating of the ochre outdoors under a full moon (Lewis-Williams 2001: 30).

Mineral biographies

After their production as artefacts and monuments, minerals continue to be drawn into symbolic associations and socially meaningful relationships over the course of their use. This is clearly illustrated by the fact that ethnographic and ethnohistorical records indicate that many stone tools have uses that are partly or wholly ritual in nature (Brumm, this volume; Cane 1988; Skeates 1995; Taçon 1991; Warburton and Duke 1995), and in some cases tools have been produced less for use than for trade within ritualised exchange systems that serve to articulate social relations and obligations (Brumm, this volume; Hampton 1999;

Jones and White 1988; McBryde 1997a, 1997b, 2000; Strathern 1969). In the recent ethnographic past and still to some degree in the present, Aboriginal Australians have exchanged stone implements within trade networks that extend over vast distances and are described as being amongst the most extensive known from hunter-gatherer societies worldwide (McBryde 1997a; see also Jones and White 1988). Paton has argued that *leilira* blades exchanged within such networks in the Northern Territory of Australia have little intrinsic value either as utilitarian items or raw materials which could be used to manufacture more functional tools (Paton 1994). Instead, he argues, the real value of these artefacts lies in the socially indispensable messages they are used to communicate. In his chapter, Brumm argues that the symbolic values attached to such stone implements, and in particular the ubiquitous stone axe, have some antiquity in Australia, and can potentially even be linked to the historical origins of the Dreaming. Systems of exchange involving symbolically-meaningful stone axes have also been proposed for the Neolithic in certain regions of Europe (Bradley 1990; Bradley and Edmonds 1993; Patton 1991; Skeates 1995; Thomas 1996; Whittle 1995). Studies have shown that axes were traded over substantial distances, and were made from visually distinctive stone sources that would have facilitated the recognition of sources (Cooney 1998, 2002; Thomas 1996).

The notion of artefact 'biographies' suggests that it is important to look at the use of mineral objects and monuments in temporal perspective. Owoc's work, for example, stresses that Bronze Age funerary monuments must be understood in terms of the shifts that they undergo through time (Owoc forthcoming, and this volume). As Owoc describes in this volume, monuments became 'harder' over time, for example as the use of looser or more friable sedimentary fabrics was superseded by the employment of more compact and brittle materials. Such changes likely signalled social transitions, and changes in the relationship between the living and the dead. A very different example from South Africa suggests that rock art should also usefully be examined in temporal perspective. Ouzman argues that rock surfaces were understood by San peoples of South Africa as permeable interfaces between Ordinary and Spirit worlds, so that rock engraving, which removed the outer layer of rock to expose the lighter, contrasting rock beneath, exposed the Spirit World, allowing it to be glimpsed by the Ordinary (Ouzman 2001). Weathering processes that dulled rock engravings with a dark patina thus represented the reclaiming of engraved images or beings by the Spirit World, and various processes of 'renewal' involving the hammering and rubbing of images were therefore needed to maintain contact between the two worlds (*ibid*).

Jones has extensively discussed the importance of artefact biographies, particularly in the construction of personhood, and many of his examples are drawn from the mineral world (Jones 2002a, 2002b, forthcoming; see also Thomas 1996, 1999). He notes, in particular, the importance of processes of exchange and curation in embodying memories as well as relationships amongst the living. Jones' discussion of fragmentation and deposition is also relevant:

composite mineral artefacts such as necklaces of jet beads may be broken up and aggregated in everyday and burial contexts in order to establish and negotiate relationships between individuals and communities (Jones 2002a, forthcoming; see also Sheridan and Davis 1998 for a discussion of jet object curation). Indeed, Jones' discussion highlights the fact that deposition or destruction is an often critical aspect of an object's biography (see also Brumm, this volume; Chapman 2000; Larsson 2000; Pollard 2001; Skeates 1995; Thomas 1996).

Others have similarly noted the importance of object deposition, particularly in the case of mineral objects, where deposition into the ground may be understood as a means of returning minerals to their natural state or place. Cooney (1998), for example, has found clear evidence that material associated with the working of stone during the Neolithic period in Ireland was deliberately placed back into the earth during the course of activities that likely involved both placation of and dedication to the earth. Thomas (1999) has similarly argued that Neolithic period flint mine shafts in England were deliberately infilled during the course of ritualised activities that emphasised reciprocal relationships between people and the earth. Stone axes deposited in rock fissures, and carved chalk objects deposited in pits or shafts excavated into the chalk itself may represent related types of activities (Bradley 2000; see also Thomas 1999).

Parallels can also be drawn to practices associated with terracotta sculptures in contemporary India, most of which are intended to disintegrate and dissolve (Huyler 1995). The sculptures, some of which at five metres in height are amongst the largest terracotta sculptures ever made, are created for rituals, during which they are believed to be infused with divine life. Afterwards, they are deposited in wells or lakes, or left at the edges of villages to disintegrate, in a process of clay recycling that is seen as paralleling the processes of birth, death and rebirth undergone by humans (*ibid*).

MINERALS, MATERIALITY AND THE CRITIQUE OF REPRESENTATION

What much of the foregoing discussion has made clear, then, is that minerals symbolise. What is perhaps equally clear, however, is that minerals are not abstract symbols, like the words in texts and spoken languages, arbitrarily related to the concepts that they stand for. Rather, mineral meanings emerge from the characteristics of the material mineral world, and the properties that are imminent in embodied human engagement with that world (see also Boivin forthcoming). Earth stands as a symbol for female because it is fertile and life-giving, like women. Stone is male because it is hard like penises, and penetrates the earth. Stone and bone are conceptually linked in webs of symbolic meaning because they share physical properties of hardness and durability. Ochre is compared to blood because it is red in colour, and when mixed with water may closely resemble the appearance of blood spilled by human activities. Mica may be linked to light and related notions of good and the sacred because of its sparkly quality, while quartz is meaningful because it possesses the ability to glow when rubbed or struck.

The recognition of the non-arbitrariness of material symbols has emerged gradually over the last decade (eg Boivin forthcoming; Parker Pearson and Ramilisonina 1998; Thomas 1996), and is closely linked to an increased interest in the substantive, physical properties of the material world that has drawn attention to minerals and other substances. The study of minerals, and in particular the interest they have increasingly attracted from archaeologists and others, may be seen as part of a general move away from the textual model that dominated earlier post-processual conceptualisations of material culture. While this model emphasised the significatory properties of material culture, more recent theoretical approaches have often drawn attention to the non-representative qualities of the material world (eg Bloch 1995a; Boivin forthcoming; Buchli 1995; Graves-Brown 2000; Olsen 2003). Studies of minerals have been drawn into and indeed at times encouraged such approaches. Such studies have clearly demonstrated that on many occasions, minerals have not just *represented*, but instead have actively *embodied* experiences, memories, and supernatural qualities. Indeed, such a distinction is clear in Taçon's pivotal essay on 'the power of stone' (1991). While stone may represent, it also has a power that goes beyond representation, and constitutes in some cases the kind of 'life-force' that Taçon discusses with respect to ochre in his chapter in this volume (see also Brumm, this volume; Tilley 1994, 1996).

McBryde is also clear in drawing this distinction with respect to Australian Aboriginal material landscapes. She observes that:

> It is no hard task to regard the features of the landscape as representational, symbolic of the presence of ancestral beings from the Dreaming. To move beyond this to accept these features as embodying the actual living presence of spiritual power and involving embodied, direct interaction with this makes demands of a different dimension. Yet the move is essential to understanding. Even if the encounter cannot be shared fully by those of other traditions, its cultural importance can be recognised and respected. (McBryde 2000: 156)

McBryde stresses, therefore, that certain ochres can actively *bestow* ancestral power, and not just represent it. Similar arguments are found in discussions of minerals in other cultural contexts. Postgate emphasises that the supernatural powers attributed to minerals by ancient Mesopotamians were considered to be just as 'real' as their natural properties, and were not conceptually distinguished from them (Postgate 1997). Robinson, in this volume, argues that certain minerals were considered by the Chumash to possess special qualities that went beyond the symbolic. These were associated with the presence of *'atiswin*, a complex concept having no direct analogue in English, but seeming to stand for both supernatural beings and the quality of power they embody.

Belief in the real presence of a supernatural power is apparent in discussions of the various unmodified natural stones that are collected, traded, hidden and deposited in graves and sacred places in a range of cultures. For example, the Dani of Irian Jaya collect small rocks and river worn pebbles of unusual colour or shape, believing them, like quartz, to be *wusa*, or sacred. Quartz is considered to possess spiritual or magical properties in many societies (as discussed by Saunders and Scarre in this volume; see also Whitley *et al* 1999), and there is good evidence that this was the case in the past as well. Unmodified quartz clasts have been found deliberately deposited in boulder cracks at prehistoric rock art sites in California (*ibid*), and at prehistoric monuments in England (Bradley 2000; Darvill 2002; Tilley 1996). That simple, unmodified stones have long possessed magical powers is also attested in Mesopotamian texts, which indicate that groups of mainly semi-precious stones enclosed together in a bag were commonly tied around a patient's neck as phylacteries in ancient Mesopotamia (Postgate 1997).

This last example demonstrates that the link between bodies and minerals is more than just representational. Not only can minerals stand for bodies or body parts, but they can actually actively transform, heal or protect bodies, as attested by the common use of minerals in indigenous healing practices. Amongst the Otomi of Mexico, for example, crystal is believed to pull sickness out of the body (Brady and Prufer 1999), while Saunders notes that ethnohistorical sources discuss the ingestion of clay by various South American culture groups in order to prevent disease (this volume). He also observes that in Mesoamerica, obsidian powder was applied to the eyes of those suffering from cataracts (Saunders 2001a). In a similar vein, Taçon's chapter describes how Aboriginal Australians consider the act of applying ochre to the body to be a transformative practice, in which people recharge themselves with ancestral power. Ochre has not uncommonly been used as a medicine in a wide range of cultures (Lewis-Williams 2001; Velo and Kehoe 1990).

The body also emerges as key in recent studies that foreground the senses. Again, such studies constitute a shift away from representation, in that they stress the embodied, sensual and emotive experience of the world (Csordas 1990, 1994; Jackson 1981). Notable amongst these have been studies of rock art that have moved away from a strict focus on imagery and symbolism in order to explore the materiality of stone and mineral pigments. A number of recent studies have, for example, stressed the forms, textures, and cracks featured on rock surfaces (eg Arsenault 2004; Bradley *et al* 2002; Keyser and Poetschat 2004; Lewis-Williams and Dowson 1990; Rozwadowski 2001; Taçon *et al* 1997; Whitley 1998; Whitley *et al* 1999), as well as the sounds that rocks make (eg Boivin 2004; Hedges 1993; Ouzman 2001; Rainbird 2002). Ethnography informs us that such features are often related to essences, spirit beings or spirit worlds believed to exist within or beyond the rock itself (eg Lewis-Williams 2001; Lewis-Williams and Dowson 1990; Whitley *et al* 1999). In some cases, rock art may be seen as a reflection, embodiment or creation of such beings or essences, while in others the application of pigment may be understood as a way of interacting with and accessing some of the power that these essences or beings possess (eg Ouzman 2001; Robinson, this volume; Whitley 1998). Robinson's chapter overlaps with these studies to some degree, in that it stresses the action of pigment application over imagery, and its argument that Native American philosophies emphasise sentience over mind is an important one. In Robinson's chapter, physical contact with the rock surface is paramount, and its significance is experiential and immediate, rather than abstract and symbolic. Robinson's comparison of rock art creation with dance is apt, and illustrates his concern with bodily movement and performance over representation.

Other papers in this volume also emphasise non-visual aspects of the mineral world. Scarre's chapter features an extensive discussion of the texture of stones used to construct Neolithic megalithic monuments in north-western France (see also Cummings 2002; MacGregor 2002), while Owoc's addresses both the visual and tactile qualities of mineral substances incorporated into Bronze Age funerary monuments in south-west Britain (see also Owoc 2002). The chapter by Saunders, meanwhile, touches on sound, describing the use of minerals to make music amongst Amerindians. Saunders' interest in the multi-sensorial qualities of Mesoamerican and South American cosmologies features in a number of his papers (Saunders 2001a, 2002, this volume), and is closely linked to an associated interest in aesthetics. Indeed, aesthetics, as 'the science which treats of the conditions of sensuous perception' (Shelton 1992: 209), or 'the effect of the physical properties of objects on the senses and qualitative evaluation of those properties' (Morphy 1994: 258), is by definition closely linked to the senses, and has accordingly emerged as an area of interest in often direct association with the latter.

The attention that has recently been paid to the brilliant or shimmering qualities of certain materials highlights one area where aesthetics and minerals clearly overlap. Interest in the shiny qualities of particular materials and items is a prominent feature of the Australian Aboriginal cultural aesthetic, according to which brilliance is viewed as a manifestation of Ancestral power (McBryde 2000; Morphy 1989; Taçon 1991, this volume). Taçon's chapter discusses this quality in relation to ochre, and argues that an interest in things iridescent and bright can be traced back at least 6,000 years in the Australian archaeological record (see also Taçon 1991).

Archaeological and ethnohistorical studies of minerals in other parts of the world have similarly drawn attention to the potential importance of the shiny qualities of certain objects. Sillar, for example, argues that the choice of micaceous clays and the burnishing of pots evident in pre-Hispanic ceramics in the Andes might relate to a fundamental Andean belief in the divine animation of all material things (Sillar 1996). Saunders similarly argues that perceptions and uses of obsidian in pre-Columbian Mesoamerica, particularly in relation to acts of ritual bloodletting and human sacrifice, were linked to an aesthetic that valued brilliance and light, and accorded distinctive forms of agency to objects that possessed such qualities (Saunders 2001a). He observes that this aesthetic was subsequently used to subvert Spanish attempts to impose Christian monotheism, through the incorporation of obsidian into Christian ritual objects and structures (*ibid*). Saunders (2001a, 2001b, 2002) and others (Gosden 2001;

Jones 2002a) draw upon Gell's (1992) notion of 'technologies of enchantment' to address the ways that technologies of mineral artefact production often bestow agency upon artefacts through the achievement of particular aesthetic effects (see also Gell 1998). Cooney (2002), for example, observes that the grinding and polishing of Neolithic stone axes in Ireland was likely aimed in part at drawing attention to speckles within the stone, which appears to have been targeted predominantly for its ability to produce distinctive, perhaps shimmering axes.

CONCLUSION

Overall, then, the critique of representation in archaeology argues that it is just as important to consider what minerals do as to consider what they mean. Minerals, and the objects and landscapes they create, have physical, sensual, emotional, physiological and spiritual effects on people that the dominant linguistic paradigm in the social sciences simply does not address. Minerals are evocative and powerful for reasons that go beyond their mere symbolic value, and have much to do with the way that we as humans physically engage with and transform them. Archaeologists will certainly need to address such facets of human interaction with the mineral world if they hope to do more than impose modern biases and academic trends in the interpretation of a predominantly mineral-made archaeological record.

So too, it is clear, will anthropologists, cultural resource management professionals, politicians and others need to address alternative conceptualisations of the mineral world. As industrial capitalism relentlessly pursues the discovery and exploitation of ever more mineral and other natural resources, contrasting perspectives on the mineral world are brought into increasingly sharp relief (see also Frodeman, this volume; Poole 1986; Stewart and Strathern 2002). McBryde, for example, contextualises her extensive discussion of Aboriginal Australian ochre expeditions and sources within the framework of a long-term conflict between indigenous and capitalist colonial values that carries through to the present day (McBryde 2000; see also McBryde 1997a). Contradictory understandings of the mineral world were foregrounded in the early 1900s in Australia, when European mineral exploitation interests clashed with Aboriginal understandings of the sacredness of certain mineral deposits. Such tensions carry through to the present day, emerging in conflicting expectations of what exactly constitutes a heritage resource. Valuations of mineral sources and the invisible lines of travel that link them within a landscape of shared Aboriginal meanings challenge traditional Western heritage resource management approaches, and their strict focus on clearly bounded sites and characteristic polarisation of sacred and secular (see discussion in McBryde 2000). Similar, albeit less formalised conflicts are also taking place in the central Deccan region of south India at the present time, where intensive industrial-scale quarrying of an archaeologically, aesthetically and culturally distinctive granitic landscape is being contested by a diverse range of local, regional and international groups (archaeological facets of this conflict are addressed in Anonymous 2003; Boivin 2004).

The genuinely incompatible understandings that different groups today hold about the mineral and indeed material world as a whole are perhaps most effectively illustrated in one final example I wish to highlight. It is the description of a dream that a Yolngu elder named Burrumarra had, at a time when non-Aboriginal mining companies had secured consent to explore for minerals in the vicinity of Dholtji outstation in Arnhem Land. McIntosh records Burrumarra's dream as follows:

> In the dream Burrumarra's younger brothers Liwukang and Wulanybuma were clearing land for an airstrip at this ceremonial center for Birrinydji [a noted Ancestral Being] ... The brothers had nearly finished the airstrip when their bulldozer was halted by an obstacle. It was a huge gold nugget. "This must belong to Birrinydji," they thought and went off to get their older brother. The Warramiri leader stared at the find and understood that the wealth of whites could be theirs once again, and he reflected on the past. Birrinydji was the rich minerals of the earth, the transformed hematite, the source of the technology that made foreigners wealthy, allowing them to dominate Aborigines. And he bent down to pick up the prize, and as he lifted it and held it in his arms, Birrinydji, Burrumarra's Aboriginality and his Dreaming went into the ground and out of his life. He had the wealth of the Balanda [whites] and that was all that he had. The Yolngu interpreted his dream to mean that to savage the earth for its "spirit," as in mining, was to lose one's identity and become like the Balanda. By resisting the temptation one would maintain one's Aboriginality but also one's poverty. (McIntosh 2000: 118)

The persistence of diverse perspectives on our earth's mineral heritage into the present day emphasises the need for practitioners of not just archaeology but a whole range of disciplines to take into account the possibility of alternative understandings of the mineral world. The range of themes and approaches pursued in the chapters that follow illustrates the diversity of these understandings, and the impossibility of any single methodological or theoretical paradigm effectively encompassing the richness of thought, action and creativity that characterises human engagement with the mineral world.

Notes

1 The oppositions commonly emphasised between organic/inorganic and biological/non-biological, and drawn upon by Western scientific models in defining the term 'mineral', reflect a paradigm similar to that inherent in the animal-vegetable-mineral model. The definition of 'mineral' as 'any of the species into which inorganic substances are classified', and of 'inorganic' as 'not organic, usually of mineral origin' in the *Oxford Dictionary* reflects the interdependent and indeed mutually-defining nature of these terms. Frodeman (this volume) also problematises the organic/inorganic dichotomy.

2 Studies of material culture by sociologists are more numerous, and include much that is published in the *Journal of Material Culture*, as well as the recent studies of Daniel Miller and others (eg Miller 1995, 1998; see also Buchli and Lucas 2001).

Acknowledgments

I would like to thank Adam Brumm for providing me with some of the fascinating literature on Australian Aboriginal practices that is referenced here, and for drawing my attention to the example of Burrumarra's dream. Thanks also go to Dan Boivin for trying to help me understand the role of gold in today's economy, and to Adam Brumm, Mary Ann Owoc and Michael Petraglia for helpful editorial comments.

References

Anonymous (2003) 'Sanganakal's silent stones cry for attention', *The Hindu*, 11 April, p 3

Arsenault, D (2004) 'From natural settings to spiritual places in the Algonkian sacred landscape: an archaeological, ethnohistorical and ethnographic analysis of the Canadian shield rock art sites', in Chippindale, C and Nash, G (eds), *The Figured Landscape of Rock-Art: Looking at Pictures in Place*, Cambridge: CUP

Ashmore, W and Knapp, AB (eds) (1999) *Archaeologies of Landscape: Contemporary Perspectives*, Oxford: Blackwell

Ayyar, PVJ (1998) *South Indian Customs*, New Delhi: Rupa & Co

Bailey, DW (1990) 'The living house: signifying continuity', in Samson, R (ed), *The Social Archaeology of Houses*, Edinburgh: Edinburgh University Press

Barley, N (1994) *Smashing Pots: Feats of Clay from Africa*, London: British Museum Press

Bender, B (ed) (1993) *Landscape: Politics and Perspectives*, Oxford: Berg

Bender, B (1998) *Stonehenge: Making Space*, Oxford: Berg

Blier, SP (1987) *The Anatomy of Architecture: Ontology and Metaphor in Batammaliba Architectural Expression*, Cambridge: CUP

Bloch, M (1971) *Placing the Dead: Tombs, Ancestral Villages and Kinship Organisation in Madagascar*, London: Seminar Press

Bloch, M (1995a) 'Questions not to ask of Malagasy carvings', in Hodder, I, Shanks, M, Buchli, V, Carman, J, Last, J and Lucas, G (eds), *Interpreting Archaeology: Finding Meaning in the Past*, London: Routledge

Bloch, M (1995b) 'The resurrection of the house amongst the Zafimaniry of Madagascar', in Carsten, J and Hugh-Jones, S (eds), *About the House: Levi-Strauss and Beyond*, Cambridge: CUP

Boivin, N (2001) *'Archaeological Science as Anthropology': Time, Space and Materiality in Rural India and the Ancient Past*, unpublished PhD thesis, Department of Archaeology, University of Cambridge

Boivin, N (2004) 'Rock art and rock music: petroglyphs of the South Indian Neolithic' 78(229) *Antiquity* 38–53

Boivin, N (forthcoming) 'Mind over matter? Collapsing the mind-matter dichotomy in material culture studies', in DeMarrais, E, Renfrew, C and Gosden, C (eds), *Rethinking Materiality: The Engagement of Mind with the Material World*, Cambridge: McDonald Institute Monograph Series

Bradley, R (1990) 'Perforated stone axe-heads in the British Neolithic: their distribution and significance' 9 *Oxford Journal of Archaeology* 299–304

Bradley, R (1993) *Altering the Earth: The Origins of Monuments in Britain and Continental Europe*, Edinburgh: Society of Antiquaries of Scotland

Bradley, R (1998) *The Significance of Monuments: On the Shaping of Human Experience in Neolithic and Bronze Age Europe*, London: Routledge

Bradley, R (2000) *An Archaeology of Natural Places*, London: Routledge

Bradley, R and Edmonds, M (1993) *Interpreting the Axe Trade: Production and Exchange in Neolithic Britain*, Cambridge: CUP

Bradley, R, Jones, A, Nordenborg Myhre, L, and Sackett, H (2002) 'Sailing through stone: carved ships and the rock face at Revheim, Southwest Norway' 35(2) *Norwegian Archaeological Review* 109–18

Bradley, R, Meredith, P, Smith, J and Edmonds, M (1992) 'Rock physics and the Neolithic axe trade in Great Britain' 34(2) *Archaeometry* 223–33

Brady, JE and Prufer, KM (1999) 'Caves and crystalmancy: evidence for the use of crystals in ancient Maya religion' 55 *Journal of Anthropological Research* 129–44

Brady, JE, Scott, A, Neff, H and Glascock, MD (1997) 'Speleothem breakage, movement, removal and caching: an aspect of ancient Maya cave modification' 12(6) *Geoarchaeology* 725–50

Buchli, VA (1995) 'Interpreting material culture: the trouble with text', in Hodder, I, Shanks, M, Buchli, V, Carman, J, Last, J and Lucas, G (eds), *Interpreting Archaeology: Finding Meaning in the Past*, London: Routledge

Buchli, VA and Lucas, G (2001) *The Archaeologies of the Contemporary Past*, London: Routledge

Bukach, D (2003) 'Exploring identity and place: an analysis of the provenance of passage grave stones on Guernsey and Jersey in the Middle Neolithic' 22(1) *Oxford Journal of Archaeology* 23–33

Burton, J (1984) 'Quarrying in a tribal society' 16(2) *World Archaeology* 234–47

Cane, SB (1988) 'Written on stone: a discussion on ethnographic and Aboriginal perspection of stone tools', in Meehan, B and Jones, R (eds), *Archaeology with Ethnography: An Australian Perspective*, Canberra: Department of Prehistory, Research School of Pacific Studies, Australian National University

Chapman, J (1994) 'The living, the dead and the ancestors: time, life cycles and the mortuary domain in later European prehistory', in Davies, J (ed), *Ritual and Remembrance: Responses to Death in Human Societies*, Sheffield: Sheffield Academic Press

Chapman, J (2000) *Fragmentation in Archaeology: People, Places and Broken Objects in the Prehistory of South-Eastern Europe*, London: Routledge

Childs, ST and Killick, D (1993) 'Indigenous African metallurgy: nature and culture' 22 *Annual Review of Anthropology* 317–37

Chippindale, C and Nash, G (2004) *The Figured Landscape of Rock-Art: Looking at Pictures in Place*, Cambridge: CUP

Cloke, P and Jones, O (eds) (2002) *Tree Cultures: The Place of Trees and Trees in their Place*, Oxford: Berg

Cooney, G (1998) 'Breaking stones, making places: the social landscape of axe production sites', in Gibson, A and Simpson, D (eds), *Prehistoric Ritual and Religion*, Phoenix Mill: Sutton

Cooney, G (2002) 'So many shades of rock: colour symbolism and Irish stone axeheads', in Jones, A and MacGregor, G (eds), *Colouring the Past*, Oxford: Berg

Coote, J and Shelton, A (eds) (1992) *Anthropology, Art and Aesthetics*, Oxford: Clarendon

Counihan, C and Van Esterik, P (eds) (1997) *Food and Culture: A Reader*, London: Routledge

Csordas, TJ (1990) 'Embodiment as a paradigm for anthropology' 18 *Ethos* 5–47

Csordas, TJ (ed) (1994) *Embodiment and Experience: The Existential Ground of Culture and Self*, Cambridge: CUP

Cummings, V (2002) 'Experiencing texture and transformation in the British Neolithic' 21(3) *Oxford Journal of Archaeology* 249–61

Darvill, T (2002) 'White on blonde: quartz pebbles and the use of quartz at Neolithic monuments in the Isle of Man and beyond', in Jones, A and MacGregor, G (eds), *Colouring the Past*, Oxford: Berg

Descola, P and Palsson, G (eds) (1996) *Nature and Society: Anthropological Perspectives*, London: Routledge

Dietler, M and Herbich, I (1998) 'Habitus, techniques, style: an integrated approach to the social understanding of material culture and boundaries', in Stark, MT (ed), *The Archaeology of Social Boundaries*, Washington, DC: Smithsonian Institution Press

Dobres, M-A and Hoffman, CR (eds) (1999) *The Social Dynamics of Technology: Practice, Politics and World Views*, Washington, DC: Smithsonian Institution Press

Ellen, R and Fukui, K (eds) (1996) *Redefining Nature: Ecology, Culture and Domination*, Oxford: Berg

Faris, JC (1972) *Nuba Personal Art*, London: Duckworth

Gell, A (1992) 'The technology of enchantment and the enchantment of technology', in Coote, J and Shelton, A (eds), *Anthropology, Art and Aesthetics*, Oxford: Clarendon

Gell, A (1998) *Art and Agency: An Anthropological Theory*, Oxford: OUP

Gosden, C (2001) 'Making sense: archaeology and aesthetics' 33(2) *World Archaeology* 163–67

Gosden, C and Hather, J (eds) (1999) *The Prehistory of Food: Appetites for Change*, London: Routledge

Gosden, C and Marshall, Y (1999) 'The cultural biography of objects' 31(2) *World Archaeology* 169–78

Gosselain, OP (1998) 'Social and technical identity in a clay crystal ball', in Stark, MT (ed), *The Archaeology of Social Boundaries*, Washington, DC: Smithsonian Institution Press

Gosselain, OP (1999) 'In pots we trust: the processing of clay and symbols in sub-Saharan Africa' 4(2) *Journal of Material Culture* 205–30

Graves-Brown, P (2000) 'Introduction', in Graves-Brown, P (ed), *Matter, Materiality and Modern Culture*, London: Routledge

Gremillion, KJ (ed) (1997) *Peoples, Plants and Landscapes: Studies in Palaeoethnobotany*, Tuscaloosa: University of Alabama Press

Haaland, G, Haaland, R and Rijal, S (2002) 'The social life of iron: a cross-cultural study of technological, symbolic and social aspects of iron making' 97(1) *Anthropos* 35–54

Hampton, OW (1999) *Culture of Stone: Sacred and Profane Uses of Stone Among the Dani*, College Station: Texas A and M University Press

Hedges, K (1993) 'Places to see and places to hear: rock art and features of the sacred landscape', in Steinbring, J, Watchman, A, Faulstich, P and Taçon, P (eds), *Time and Space: Dating and Spatial Considerations in Rock Art Research*, Occasional AURA Publication No 8, Melbourne: Australian Rock Art Research Association

Herbert, E (1993) *Iron, Gender and Power: Rituals of Transformation in African Societies*, Bloomington and Indianapolis: Indiana University Press

Hosler, D (1994) *The Sounds and the Colors of Power: The Sacred Metallurgical Technology of Ancient West Mexico*, Cambridge, MA: MIT Press

Hosler, D (1995) 'Sound, color and meaning in the metallurgy of Ancient West Mexico' 27(1) *World Archaeology* 100–15

Huntington, R and Metcalfe, P (1979) *Celebrations of Death*, Cambridge: CUP

Huyler, S (1995) 'Clay, sacred and sublime: terracotta in India', in Fisher, N (ed), *Mud, Mirror and Thread. Folk Traditions of Rural India*, Ahmedabad: Mapin

Ingold, T (ed) (1986) *The Appropriation of Nature: Essays on Human Ecology and Social Relations*, Manchester: Manchester University Press

Ingold, T (ed) (1988) *What Is An Animal?*, London: Unwin Hyman

Jackson, M (1981) 'Knowledge of the body' 18 *Man (NS)* 327–45

Jain, J (1984) *Painted Myths of Creation: Art and Ritual of an Indian Tribe*, New Delhi: Lalit Kala Akademi

Jain, J (1995) 'The implicit and the manifest in Indian folk art and mythology', in Fisher, N (ed), *Mud, Mirror and Thread: Folk Traditions of Rural India*, Ahmedabad: Mapin

Jones, A (2000) 'Life after death: monuments, material culture and social change in Neolithic Orkney', in Ritchie, A (ed), *Neolithic Orkney in its European Context*, Cambridge: McDonald Institute Monograph Series

Jones, A (2002a) 'A biography of colour: colour, material histories and personhood in the Early Bronze Age of Britain and Ireland', in Jones, A and MacGregor, G (eds), *Colouring the Past*, Oxford: Berg

Jones, A (2002b) *Archaeological Theory and Scientific Practice*, Cambridge: CUP

Jones, A (forthcoming) 'Materialising memory: colour, remembrance and the Neolithic/Bronze Age transition', in DeMarrais, E, Renfrew, C and Gosden, C (eds), *Rethinking Materiality: The Engagement of Mind with the Material World*, Cambridge: McDonald Institute Monograph Series

Jones, A and MacGregor, G (eds) (2002) *Colouring the Past*, Oxford: Berg

Jones, R and White, N (1988) 'Point blank: stone tool manufacture at the Ngilipitji Quarry, Arnhem Land, 1981', in Meehan, B and Jones, R (eds), *Archaeology with Ethnography: An Australian Perspective*, Canberra: Department of Prehistory, Research School of Pacific Studies, Australian National University

Kahn, M (1990) 'Stone-faced ancestors: the spatial anchoring of myth in Wamira, Papua New Guinea' 29 *Ethnology* 51–66

Keates, S (2002) 'The flashing blade: copper, colour and luminosity in North Italian Copper Age society', in Jones, A and MacGregor, G (eds), *Colouring the Past*, Oxford: Berg

Kennet, D and Rao, JVP (forthcoming) 'Two Early Historic brick temples at Paithan in Maharashtra', in Raven, E (ed), *South Asian Archaeology 1999. Papers of the 15th International Conference of the European Association of South Asian Archaeologists, Leiden, July 1999*

Keyser, JD and Poetschat, G (2004) 'The canvas of the art: landscape analysis of the rock art panel', in Chippindale, C and Nash, G (eds), *The Figured Landscape of Rock-Art: Looking at Pictures in Place*, Cambridge: CUP

Kinsley, D (1986) *Hindu Goddesses: Visions of the Divine Feminine in the Hindu Religious Tradition*, London: University of California Press

Kopytoff, I (1986) 'The cultural biography of things: commoditization as process', in Appadurai, A (ed), *The Social Life of Things: Commodities in Cultural Perspective*, Cambridge: CUP

Küchler, S (1985) 'Malangan: art and memory in a Melanesian society' 22(2) *Man* 238–55

Küchler, S (1992) 'Making skins: Malangan and the idiom of kinship in Northern New Ireland', in Coote, J and Shelton, A (eds), *Anthropology, Art and Aesthetics*, Oxford: Clarendon

Larsson, L (2000) 'The passage of axes: fire transformation of flint objects in the Neolithic of southern Sweden' 74 *Antiquity* 602–10

Lemonnier, P (1992) *Elements for an Anthropology of Technology*, Ann Arbor: University of Michigan Press

Levy, JE (1999) 'Metals, symbols and society in Bronze Age Denmark', in Robb, J (ed), *Material Symbols: Culture and Economy in Prehistory*, Carbondale, Ill: Center for Archaeological Investigations, Southern Illinois University

Lewis-Williams, JD (2001) 'Southern African shamanistic rock art in its social and cognitive contexts', in Price, N (ed), *The Archaeology of Shamanism*, London: Routledge

Lewis-Williams, JD and Dowson, TA (1990) 'Through the veil: San rock paintings and the rock face' 45 *South African Archaeological Bulletin* 5–16

Livingstone-Smith, A (2000) 'Processing clay for pottery in Northern Cameroon' 42(1) *Archaeometry* 21–42

Lovell, N (1998) 'Wild gods, containing wombs and moving pots: emplacement and transience in Watchi belonging', in Lovell, N (ed), *Locality and Belonging*, London: Routledge

MacGregor, G (2002) 'Making monuments out of mountains: the role of colour and texture in the constitution of meaning and identity in recumbant stone circles', in Jones, A and MacGregor, G (eds), *Colouring the Past*, Oxford: Berg

McBryde, I (1997a) 'The cultural landscapes of Aboriginal long distance exchange systems: can they be confined within our heritage registers?' 13(3–4) *Historic Environment* 6–14

McBryde, I (1997b) '"The landscape is a series of stories": grindstones, quarries and exchange in Aboriginal Australia: a Lake Eyre case study', in Ramos-Millán, A and Bustillo, MA (eds), *Siliceous Rocks and Culture*, Granada: University of Granada, Monográfica Arte y Arqueología

McBryde, I (2000) 'Travellers in storied landscapes: a case study in exchanges and heritage' 24 *Aboriginal History* 152–74

McBryde, I and Watchman, A (1976) 'The distribution of greenstone axes in south-eastern Australia: a preliminary report' 10 *Mankind* 163–74

McIntosh, IS (2000) *Aboriginal Reconciliation and the Dreaming: Warramiri Yolngu and the Quest for Equality*, Boston: Allyn and Bacon

Manning, A and Serpell, J (eds) (1994) *Animals and Human Society: Changing Perspectives*, London: Routledge

Miller, D (ed) (1995) *Acknowledging Consumption: A Review of New Studies*, London: Routledge

Miller, D (ed) (1998) *Material Cultures: Why Some Things Matter*, London: UCL Press

Morphy, H (1989) 'From dull to brilliant: the aesthetics of spiritual power among the Yolngu' 24 *Man (NS)* 21–40

Morphy, H (1991) *Ancestral Connections: Art and an Aboriginal System of Knowledge*, Chicago: University of Chicago Press

Morphy, H (1994) 'Aesthetics across time and place: an anthropological perspective' 4(2) *Cambridge Archaeological Journal* 257–60

Munn, N (1986) *Walbiri Iconography*, Chicago: University of Chicago Press

Olsen, B (2003) 'Material culture after text: re-membering things' 36(2) *Norwegian Archaeological Review* 87–104

Ouzman, S (2001) 'Seeing is deceiving: rock art and the non-visual' 33(2) *World Archaeology* 237–56

Owoc, MA (2002) 'Munselling the mound: the use of soil colour as metaphor in British Bronze Age funerary ritual', in Jones, A and MacGregor, G (eds), *Colouring the Past*, Oxford: Berg

Owoc, MA (forthcoming) 'Beyond geoarchaeology: pragmatist explorations of an alternative British Bronze Age viewscape', in Robertson, J, Siebert, J, Fernandes, D and Zender, M (eds), *Space and Spatial Analysis in Archaeology*, Calgary: University of Calgary Press

Parker Pearson, M (2000) 'Ancestors, bones and stones in Neolithic and Early Iron Age Britain and Ireland', in Ritchie, A (ed), *Neolithic Orkney in its European Context*, Cambridge: McDonald Institute for Archaeological Research

Parker Pearson, M (2002) 'Placing the physical and incorporeal dead: Stonehenge and changing concepts of ancestral space in Neolithic Britain', in Silverman, H and Small, DB (eds), *The Space and Place of Death*, Archaeological Papers of the American Anthropological Association No 11

Parker Pearson, M and Ramilisonina (1998) 'Stonehenge for the ancestors: the stones pass on the message' 72 *Antiquity* 308–26

Paton, R (1994) 'Speaking through stones: a study from northern Australia' 26(2) *World Archaeology* 172–84

Patton, MA (1991) 'Axes, men and women: symbolic dimensions of Neolithic exchange in Armorica (north-west France)', in Garwood, P, Jennings, D, Skeates, R and Toms, J (eds), *Sacred and Profane: Proceedings of a Conference on Archaeology, Ritual and Religion*, Oxford: Oxford Committee for Archaeology

Patton, MA (1993) *Statements in Stone: Monuments and Society in Neolithic Brittany*, London: Routledge

Petraglia, MD, Bupp, SL, Fitzell, SP and Cunningham, KW (2002) *Hickory Bluff: Changing Perceptions of Delmarva Archaeology*, Dover: Delaware Department of Transportation Archaeology Series

Pfaffenberger, B (1988) 'Fetishised objects and humanised culture: towards an anthropology of technology' 23 *Man (NS)* 236–52

Pfaffenberger, B (1992) 'Social anthropology of technology' 21 *Annual Review of Anthropology* 491–516

Pollard, J (2001) 'The aesthetics of depositional practice' 33(2) *World Archaeology* 315–33

Poole, F (1986) 'The erosion of a sacred landscape: European exploration and cultural ecology among the Bimin-Kuskusmin of Papua New Guinea', in Tobias, M (ed), *Mountain People*, Oklahoma: University of Oklahoma Press

Postgate, N (1997) 'Mesopotamian petrology: stages in the classification of the material world' 7(2) *Cambridge Archaeological Journal* 205–24

Rainbird, P (2002) 'Making sense of petroglyphs: the sound of rock-art', in David, B and Wilson, M (eds), *Inscribed Landscapes: Marking and Making Place*, Honolulu: University of Hawaii Press

Renne, EP (1991) 'Water, spirits and plain white cloth: the ambiguity of things in Bunu social life' 26 *Man (NS)* 709–22

Richards, C (1996) 'Henges and water: towards an elemental understanding of monumentality and landscape in Late Neolithic Britain' 1 *Journal of Material Culture* 313–36

Rival, L (ed) (1998) *The Social Life of Trees: Anthropological Perspectives on Tree Symbolism*, Oxford: Berg

Roe, D and Taki, J (1999) 'Living with stones: people and the landscape in Erromango, Vanuatu', in Ucko, PJ and Layton, R (eds), *The Archaeology and Anthropology of Landscape*, London: Routledge

Rozwadowski, A (2001) 'Sun gods or shamans? Interpreting the 'solar-headed' petroglyphs of Central Asia', in Price, N (ed), *The Archaeology of Shamanism*, London: Routledge

Saunders, NJ (2001a) 'A dark light: reflections on obsidian in Mesoamerica' 33(2) *World Archaeology* 220–36

Saunders, NJ (2001b) '"Catching the light": technologies of power and enchantment in pre-Columbian goldworking', in Quilter, J and Hoopes, JW (eds), *Gold and Power in Ancient Costa Rica, Panama, and Columbia*, Washington, DC: Dumbarton Oaks Research Library and Collection

Saunders, NJ (2002) 'The colours of light: materiality and chromatic cultures of the Americas', in Jones, A and MacGregor, G (eds), *Colouring the Past*, Oxford: Berg

Scarre, C (ed) (2002) *Monuments and Landscapes in Atlantic Europe*, London: Routledge

Schmidt, PR (1997) *Iron Technology in East Africa: Symbolism, Science and Archaeology*, Bloomington: Indiana University Press

Shelton, A (1992) 'Predicates of aesthetic judgement: ontology and value in Huichol material representations', in Coote, J and Shelton, A (eds), *Anthropology, Art and Aesthetics*, Oxford: Clarendon

Sheridan, A and Davis, M (1998) 'The Welsh "jet set" in prehistory: a case of keeping up with the Joneses?', in Gibson, A and Simpson, D (eds), *Prehistoric Ritual and Religion*, Phoenix Mill: Sutton

Sillar, B (1996) 'The dead and the drying: techniques for transforming people and things in the Andes' 1(3) *Journal of Material Culture* 259–89

Skeates, R (1995) 'Animate objects: a biography of prehistoric axe-amulets in the central Mediterranean region' 61 *Proceedings of the Prehistoric Society* 279–301

Spence, K (1999) 'Red, white and black: colour in building stone in ancient Egypt' 9(1) *Cambridge Archaeological Journal* 114–17

Stewart, P and Strathern, A (2002) *Remaking the World: Myth, Mining and Ritual Change Among the Duna of Papua New Guinea*, Washington, DC: Smithsonian Institution Press

Strathern, M (1969) 'Stone axes and flake tools: evaluations from two New Guinea highland societies' 35 *Proceedings of the Prehistoric Society* 311–29

Strauss, S and Orlove, BS (eds) (2003) *Weather, Climate, Culture*, Oxford: Berg

Taçon, PSC (1990) 'The power of place: cross-cultural responses to natural and cultural landscapes of stone and earth', in Vastokas, JM, Paper, J and Taçon, PSC (eds), *Perspectives of Canadian Landscape: Native Traditions*, York: Robarts Centre for Canadian Studies, York University

Taçon, PSC (1991) 'The power of stone: symbolic aspects of stone use and tool development in western Arnhem Land, Australia' 65 *Antiquity* 192–207

Taçon, PSC and Ouzman, S (2004) 'Worlds within stone: the inner and outer rock-art landscapes of northern Australia and southern Africa', in Chippindale, C and Nash, G (eds), *The Figured Landscape of Rock-Art: Looking at Pictures in Place*, Cambridge: CUP

Taçon, PSC, Fullager, R, Ouzman, S and Mulvaney, K (1997) 'Cupule engravings from Jinmium-Granilpi (northern Australia) and beyond: exploration of a widespread and enigmatic class of rock markings' 71 *Antiquity* 942–65

Thomas, J (1996) *Time, Culture and Identity: An Interpretive Archaeology*, London: Routledge

Thomas, J (1999) 'An economy of substances in earlier Neolithic Britain', in Robb, J (ed), *Material Symbols: Culture and Economy in Prehistory*, Carbondale, Illinois: Centre for Archaeological Investigations, Southern Illinois University

Tilley, C (1994) *A Phenomenology of Landscape*, Oxford: Berg

Tilley, C (1996) 'The powers of rocks: topography and monument construction on Bodmin Moor' 28(2) *World Archaeology* 161–76

Ucko, PJ and Layton, R (eds) (1999) *Archaeology and Anthropology of Landscape: Shaping Your Landscape*, London: Routledge

Van Nest, J, Charles, DK, Buikstra, JE and Asch, DL (2001) 'Sod blocks in Illinois Hopewell mounds' 66(4) *American Antiquity* 633–50

Velo, J and Kehoe, AB (1990) 'Red ochre in the Paleolithic', in Foster, ML and Botscharow, LJ (eds), *The Life of Symbols*, Oxford: Westview

Warburton, M and Duke, P (1995) 'Projectile points as cultural symbols: ethnography and archaeology', in Duke, P and Wilson, MC (eds), *Beyond Subsistence: Plains Archaeology and the Post-Processual Critique*, Tuscaloosa: University of Alabama Press

Watson, C (2003) *Piercing the Ground*, Fremantle: Fremantle Arts Centre Press

Whitley, DS (1998) 'Finding rain in the desert: landscape, gender and far western North American rock-art', in Chippindale, C and Taçon, PSC (eds), *The Archaeology of Rock-Art*, Cambridge: CUP

Whitley, DS, Dorn, RI, Simon, JM, Rechtman, R and Whitley, TK (1999) 'Sally's rockshelter and the archaeology of the Vision Quest' 9(2) *Cambridge Archaeological Journal* 221–47

Whittle, A (1995) 'Gifts from the earth: symbolic dimensions of the use and production of Neolithic flint and stone axes' 33 *Archaeologia Polona* 247–59

Wills, R (ed) (1990) *Signifying Animals: Human Meaning in the Natural World*, London: Unwin Hyman

– 2 –

OCHRE, CLAY, STONE AND ART

THE SYMBOLIC IMPORTANCE OF MINERALS AS LIFE-FORCE AMONG ABORIGINAL PEOPLES OF NORTHERN AND CENTRAL AUSTRALIA

Paul SC Taçon

ANTIQUITY OF OCHRE, CLAY AND EARTH AS SYMBOLIC SUBSTANCES

In industrialised countries and many other parts of the world, minerals are often thought of as commodities, substances from the earth to be bought, sold, made into objects and accumulated for profit. However, haematite, crystal, gold, silver, diamonds and many other minerals have long had important symbolic roles worldwide. Indeed, things that are bright and colourful are often especially potent (see eg Clark 1986; Taçon 1999). Among the Aboriginal peoples of northern Australia this is also true, brightness and colour being associated with the power of great Ancestral Beings. In this paper, a brief overview of the ways in which ochres, clay and other earthy substances are used by recent and contemporary Aboriginal peoples in Australia is undertaken with a view to emphasising that they play essential symbolic roles at every stage of life.

The late Big Bill Neidjie, 'Kakadu Man' to many people, was both a poet and a philosopher. Among other things, he was particularly proud of his culture and heritage, including the traditional artistic and ceremonial practices that helped to define them. In the 1980s, he was one of my teachers, helping to initiate me into the multi-layered nature of Aboriginal society, meaning and existence. On many occasions he would talk about the importance of earth by emphasising our connections to it: 'We come from earth ... bones, We go to earth ... ashes', he would say (as in Neidjie *et al* 1985: 46). In publications of his knowledge and wisdom (*ibid*; Neidjie 1989) he often emphasised this connection. For instance:

> Earth ...
> like your father or brother or mother,
> because you born from earth.
> You got to come back to earth.
> When you dead ...
> you'll come back to earth.
> Maybe little while yet ...

then you'll come to earth.
That's your bone,
your blood.
It's in this earth,
same as for tree. (Neidjie *et al* 1985: 51)

One of the most important pigments derived from the earth is ochre, used by Aboriginal peoples in every significant ceremony, from birth to death. Indeed, as Mulvaney (1978: 15) noted some time ago, '[o]chre is a recurring archaeological theme', for all periods in Australia. According to Smith (2000: 665), some of the more impressive ochre quarry sites, including 'Bookartoo (SA), Wilgie Mia (WA), Karrku (NT), and Toolumbunner (Tas)', were used for hundreds, if not thousands, of years. The use of ochre was integral to the first people to reach Australia at least 43,000 to 65,000 years ago, as evidenced by used pieces of ochre in the lowest levels of Australia's oldest sites. At Malakunanja II and Nauwalabila in Kakadu National Park, Northern Territory, for instance, the lowest levels of which OSL date to 55000–60000 BP, used pieces of ochre are found throughout the deposits (Roberts *et al* 1990, 1993, 1998). In the north of Western Australia, at Carpenter's Gap and other sites, fragments of red ochre with use striations have also been found at the deepest levels. Also found at Carpenter's Gap was a piece of the oldest surviving and securely dated rock painting in Australia. The ochre-painted rock fragment appears to have fallen from an old ceiling and was recovered from a layer radiocarbon-dated to 39000 BP (O'Connor 1995). However, because of its fragmentary nature, what it originally partly depicted is undecipherable. Elsewhere, in Western Australia and in Australia's central region, the lowest levels of a range of sites have been radiocarbon-dated to between 23000 and 47000 BP (eg Devil's Lair, WA [up to 47000 BP, Dortch 1979; Turney *et al* 2001], Serpent's Glen, WA [up to 23000 BP, O'Connor *et al* 1998], Puritjarra, NT [up to 32000 BP, Smith *et al* 1998, 2001]). They also contain red ochre at or near these levels, suggesting that art was practiced as soon as new areas were colonised.

Far to the south, in western New South Wales, burials of various ages have been found with red ochre. The oldest dates to close to 61000 BP using a range of dating techniques (Thorne *et al* 1999; but see Bowler *et al* 2003, who contend that the burial may be 42,000–45,000 years old). Skeletal remains of a male individual have been found to contain DNA (Adcock *et al* 2001). Importantly, there is much red ochre in the burial (Thorne 1971, 1975; Thorne *et al* 1999). As Thorne *et al* (1999: 610) note, this, combined with evidence of early ochre use from rock-shelter excavations, indicates that 'ochre was widely used for artistic and/or philosophical purposes from an early date in Australia'. They also argue that the dating results suggest that 'people who were skeletally within the range of the present Australia indigenous population colonized the continent during or before oxygen isotope stage 4 (57000–71000 BP)'.

One of the key conclusions we can reach from recent archaeological research is that the first Australians brought pigment usage with them as part of their cultural experience and knowledge. In other words, ochre use was something they and their ancestors engaged in before they arrived in Australia. A second conclusion is that from arrival, perhaps as early as 60,000 years ago, ochres were used for a range of purposes, from ritual (eg burials) to rock-art, and probably also body art. Certainly, ochre 'crayons' were commonplace, surviving in Australia's most ancient archaeological deposits across the country. This is in keeping with recent evidence from Africa that early humans were using ochre crayons as early as 77,000 years ago – although there they were also sometimes incising them with geometric designs (see Henshilwood *et al* 2002).

RECENT AND CONTEMPORARY OCHRE USAGE

As soon as Europeans arrived in Australia in the 1700s, the prominent role of ochre in Aboriginal culture was noted. Ochre used for body art, object decoration, great expanses of rock-art, and bark paintings in abandoned shelters, as well as significant ochre deposits, were all observed by early explorers. For instance, in 1705 Marten van Delft, commander of three Dutch East India ships, noted particularly that the men he had seen on Melville Island had cicatrices (raised body scars) and were adorned with 'a yellow or red ointment, prepared with turtle fat', among other things (Robert 1979: 41). In 1756, Gerrit de Haan of Batavia reported on Aboriginal people he encountered on the western coast of Cape York, Queensland, 'all of them with bodies daubed over with red' (in McKnight 1969: 43). The British also observed the use of ochre on bodies, rock and elsewhere soon after they arrived, settled and began exploring Australia in the late 1700s and early 1800s (eg see Earl 1846 for coastal Northern Territory). Matthew Flinders was particularly struck by rock paintings 'discovered' at Chasm Island on 14 January 1803 (see McCarthy 1955). Campbell (1834: 157), Carrington (1890), Wilkins (1929: 185 and plate opposite 177) and others provide first accounts of bark paintings in shelters. More generally, as Horton (1994: 820) recently surmised in the *Encyclopedia of Aboriginal Australia*, ochre was:

> the most important pigment used by Aboriginal people for painting (both for body decoration and art), staining and in ritual. Ochre is iron oxide and is found in a range of colours from yellow through red to brown. It can be found casually, but some particularly rich outcrops were mined. High-quality ochre from such mining was a valuable trade item, sometimes over considerable distances. It is used by grinding to a powder and then mixing with a fluid, usually just water but often with a binding agent such as saliva or blood. For staining skins and hair the dry powder is generally simply rubbed in.
>
> Red ochre is extremely important in desert cultures. Its great power derives from the belief that, as the metamorphosed blood of the ancestral beings, it cures, protects and strengthens.

Gage (1999: 111) also emphasises the central role of ochre in Aboriginal culture, noting for Aboriginal Australia generally that:

> In many cases terms for white, red and yellow refer to the ochres used in body and other painting, and the term for 'red' is sometimes also the term for 'blood'. The salience of 'black', 'white' and 'red' in these languages can indeed be linked to the role of these colours and pigments in painting, and they were, and are, heavily freighted with symbolism. Red ochre was regarded as especially sacred, and where there were no good local supplies, it was imported, often over long distances.

Smith (2000: 665) also notes a more practical use of ochre, 'as a preservative coating covering balls of desiccated fruit'. Mulvaney has published various accounts of the pivotal place of ochre in Aboriginal culture (eg 1969, 1978; Mulvaney and Kamminga 1999), studying its importance first hand from archaeological excavations and noting the extreme distances over which ochre was traded. For instance:

> Yarar is a painted rock shelter in the Northern Territory, occupied over a few thousand years. Excavation of part of its floor recovered 20,000 ochre fragments, weighing 183 kilograms. If this seems conspicuous waste, it is recorded that a century ago up to seventy Dieri tribesman from Cooper's Creek travelled 500 kilometres annually to collect ochre from the Flinders Range quarries and journey home with individual loads of thirty kilograms. (Mulvaney 1978: 16)

Further north, Chaloupka's (1993) ethnographic account of the preparation and use of pigment among people of western Arnhem Land is one of the most detailed available for anywhere in Australia. He describes pigment and colour terminology from indigenous and European perspectives, devoting much discussion to both practical and symbolic aspects of pigment and paint usage (see especially Chaloupka 1993: 83–86). Sources of pigment can be summarised as follows (*ibid*: 83):

> White: kaolin clays most common; also gypsum, burnt selenite, calcite, huntite (Calcium magnesium carbonate)
>
> Yellow/brown: limonite
>
> Red: haematite, including a purple of ore-grade quality, laterites, ferruginous sandstone, iron oxide stained clays; also made from burning/heating yellow ochre
>
> Black: charcoal, manganese

Pigments were collected wherever exposed, from both small and large quarries; some were mined, and large quantities were traded (**Figure 2.1 colour section**; **Figure 2.2**). The quality of pigment was very important. Essentially, the better the quality, the more potent the pigment (and, later, paint) for ceremony and other usage. As Chaloupka observes, 'although ochres and white clay can be found throughout the plateau region, the finest quality materials are usually associated with important locales and they were prized and traded over considerable distances. Their suitability is judged by their purity, consistency, opacity and texture' (Chaloupka 1993: 83).

Figure 2.2: Kakadu National Park elder David Canari gathering yellow ochre from a creek bed in northern Australia, 1990.

FIELD OBSERVATIONS ON OCHRE USE AND IMPORTANCE OF COLOUR, 1985–2002

My own first direct involvement with ochre in ceremony was at a series of funerals in Kakadu National Park in the mid-1980s. Red pigment mixed with water was routinely smeared on the upper arms of all attending in order to protect mourners from evil spirits thought to be lurking nearby. Trees near the house of the deceased and areas the deceased frequented were painted with red or yellow lines that encircled them, the colour depending on the moiety affiliation of the departed. Houses and other buildings received similar treatment. Great smears of ochre were also placed on the doors and sides of vehicles the dead person had recently been driving or riding in. All that were painted were also 'smoked' through a ceremonial wafting of burning leaves conducted by key relatives of the deceased. Sometimes people also had water poured over their heads after stepping into a special enclosure made of brightly coloured cloth supported by tree limbs and branches.[1]

In central Arnhem Land many people believe they can transform themselves by painting their bodies head to toe with white pipeclay (**Figure 2.3 colour**

section). There it is said that the white clay is the dung of the Rainbow Serpent, the most powerful Ancestral Being of all. When people paint themselves with the clay of certain creek beds they recharge themselves with ancestral power. In 1993, when I first visited a rock-art site complex near the Mann River with a large group of local people, we had to commence by digging out the white clay from the creek below. Some was bagged for later use but much was applied immediately. Soon everyone had turned from black to ghostly white, with great enthusiasm and merriment accompanying the process. We could then visit the rock-art sites with the knowledge that nothing would do us harm. It was explained that besides affording protection, painting the body was great fun. Basedow (1907: 19–20) and others report similarly for earlier periods, also noting that smearing the body with ochre and/or mud when hunting provided a protective coloured coating. Later, back at the outstation, several of our group had fun surprising their dogs, who did not recognise these strange white creatures who had replaced their owners.

Dogs themselves are actually intimately linked to pipeclay and ochre across central and northern Australia. In many areas, there are detailed stories about red and white Dingo Ancestors who travelled, fought, stole sisters and created focal points of the landscape (eg Rose 2000 [1992], 2002). At certain locations, they deposited particularly bright and powerful red or white pigment that can be used to mesmerise members of the opposite sex. Among some groups, men use red while women have a special source of white; among others this is reversed. Often these substances of earth and Ancestors are applied to portions of the face. I observed this use of ochre as recently as 2002 in north-west Queensland, among the Waanyi. More generally, Smith (2000: 665) notes that 'many red ochre deposits are associated mythologically with the blood of totemic beings – typically kangaroos, emus, dogs, or menstruating women – mirroring European use of the term "haematite"'.

Spencer and Gillen (1912: 21) describe how gypsum, 'a substance often associated with rain-men because it is white like the clouds', was ground up for use in rain-making ceremonies among peoples of central Australia. They also describe and illustrate body art associated with rain-making and other ceremonies (eg 1912: 237–43). Body art is integral to all ceremonial practice throughout Aboriginal Australia, constituting a focal point of initiation, dance and ceremony. Body art motifs and designs made on the ground or in sand are sometimes incorporated into rock-art, bark paintings, and works on canvas, paper and film. For many people, the body is made to be painted: it is a canvas created to receive the pigments of the earth, especially red, white, yellow and sometimes black, as well as shades in-between. In rock-art as old as 6,000 years through to contemporary art on bark, paper and canvas, one of the most highly decorated, colourful and brightly illuminated creatures is the Rainbow Serpent (Taçon 1989a, 1989b; Taçon *et al* 1996; Taylor 1987, 1990, 1996). Cross-hatching was used to create a 'colourful' effect from the inception of Rainbow Serpent depictions (Taçon *et al* 1996). Today it is a common feature of all subject matter in bark painting and painting on paper (see eg Morphy 1991; Taylor 1996).

Although much material culture was painted with clan, moiety and other group designs, often with elaborate cross-hatching, other important objects painted with ochre and white clay include hollow log coffins (Arnhem Land), Tiwi burial poles (Melville and Bathurst Islands) and wooden sculptures of *Mimi* spirits (Arnhem Land), Wandjina figures (Kimberley) and other beings. Sometimes human skulls were painted with intricate designs and longbones were often rubbed with red for secondary burials. The bones of macropod, emu, crocodile and other large animals were painted with red before being placed in arrangements marking a young man's first successful hunt. Sacred objects, made of bone, wood or stone, were also sometimes painted in different parts of Australia. For instance, in Arnhem Land not only was pigment and paint on stone a powerful way of expressing Ancestral potency via rock-art, but special stones were painted with designs for use in restricted men's ceremonies. These sacred objects were never shown to the uninitiated, partly because the designs were thought to be so powerful that they would bring great harm to those who did not know how to treat them properly. Indeed, on many occasions I have been told not to photograph or otherwise reveal designs painted on such objects or those on bodies painted for certain initiation ceremonies.

FROM DULL TO BRIGHT

A key aspect of painting objects and bodies involves making them not only colourful but also bright. In essence, things that are bright and colourful are considered infused with power or life-force. Crystal is one mineral that figures prominently in stories about clever men, rain-making and some initiation (eg Elkin 1930, 1944; Taçon 1989b) but other types of unusually bright and/or colourful stone were also held in high regard. For instance, the adoption of quartzite to make unifacial and bifacial stone spear points (**Figure 2.4**) and other artefacts, beginning about 6,000 years ago in western Arnhem Land, corresponds with the development of polychrome rock paintings (**Figure 2.5 colour section**). It also flags the beginning of a more intense interest in things that shimmer, are bright or exhibit iridescence (Taçon 1991). In south-east Australia, Brumm (2000; see also this volume) has documented a more recent incidence of special stone quarried, traded and revered. In this case it is green coloured stone used for making axe heads. Brumm notes that

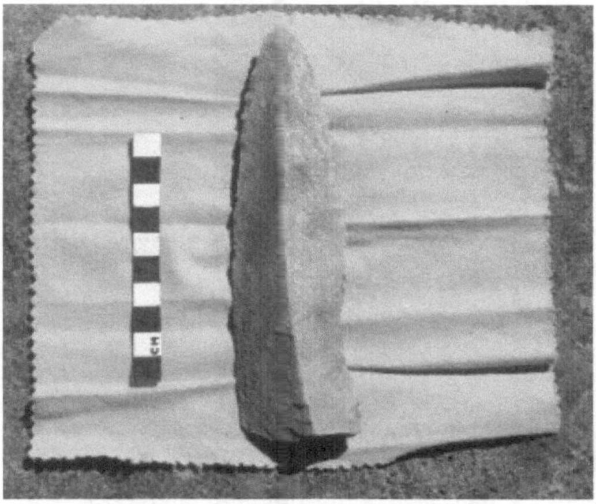

Figure 2.4: Quartzite stone spear point, Kakadu National Park.

highly polished varieties in south-east Australia may have been imbued with mystical properties in a similar way to bright/brilliant stone of the north. In both northern and southern Australia, quarries with the best bright stone were associated with powerful Ancestral Beings (Brumm 2000; Taçon 1991).

Across northern Australia, pigments derived from the land continue to be used to mark bodies, landscapes, shelters, walls, sacred objects, items of ceremony and even utilitarian material culture – to make them 'bright' with ancestral (spiritual) power. This is because, with time, all things proceed from bright to dull, necessitating renewal, repainting, re-enactment and, ultimately, reaffirmation of culture and experience, as well as expression in ever-changing new ways. Aboriginal people both mirrored and participated in this process by tapping into the power of the original Ancestral Dreamtime Beings. An understanding of this process is integral to comprehending Aboriginal metaphysics and visual art across northern Australia (see Morphy 1991, 1998; Taçon 1989a, 1989b, 1991; Taçon *et al* 1996; Taylor 1987, 1990).

An example of this worldview can be seen when Aboriginal men and women change themselves from dull to brilliant with special reds, whites and yellows from the most powerful of ancestral landscape locations in order to attract members of the opposite sex, ultimately for procreation and the birth of new generations. Bodies are also painted with ochre designs symbolic of animal species honoured in ceremony and to reaffirm social relationships during initiation. The souls of dead people are returned to the water- and earth-dwelling, spirit-child phases of existence through ritual and the action of painting the bones of the deceased with red ochre. The souls of animals are recycled, returning to ancestral wells with the wet season rains and various rituals usually involving pigment and paint. Creatures were also painted with brilliant rainbow colour on rock and/or bark shelter walls and ceilings in combinations of red, orange, purple, pink, white, cream and yellow, sometimes with black, to convey a multitude of experience and information. This form of painting continues today on sheets of bark (**Figure 2.6**), paper and other media.

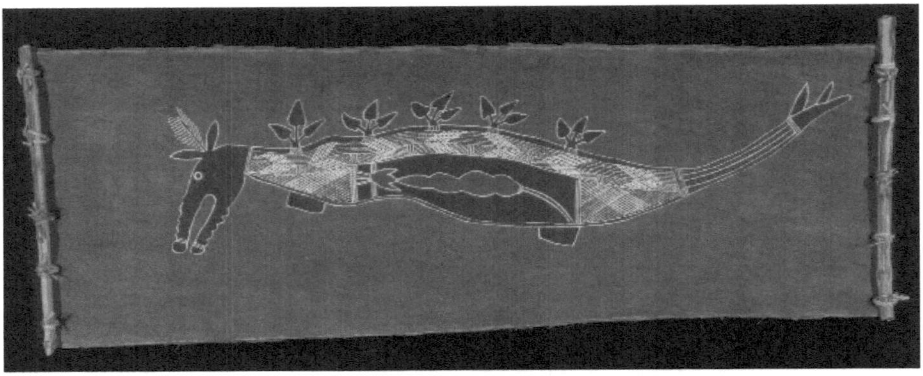

Figure 2.6: Contemporary bark painting by Bruce Nabegeyo of the 'brightest' and most powerful Ancestral Being of all, the Rainbow Serpent.

This is also one of the reasons particular animal species were repainted or superimposed over (see Sale 1995 for the most detailed paper on the re-marking of imagery). The retouching and repainting changed the shelter art back from dull to brilliant, allowing Aboriginal people to make contact with the continuing cycle of spiritual and physical existence. In the process, it reaffirmed the Aboriginal past and present for the artists and their extended families. 'In Eastern and Central Arnhem land, painting is seen as a process of transforming a surface from a state of dullness to that of shimmering brilliance (*bir'yunhamirri*)' (Morphy 1998: 188).

CONCLUSIONS

Pigments made of ochre, minerals and other substances of the earth have been widely used across Aboriginal Australia from the time of first arrivals to the present. For the Pleistocene, White and O'Connell (1982: 71) conclude that 'the widespread use of red ochre pigments, found not only in the Mungo III burial but also at many other sites, points certainly to the practice of decoration on bodies, walls or artefacts and to the possibility of ceremonial systems'. Earth pigments were used throughout the Holocene and usage appears to have increased in frequency given that excavated deposits from several locations show much more ochre left behind from the end of the Pleistocene, and especially from the mid to late Holocene (eg David *et al* 1994; Jones 1985; Taçon and Brockwell 1995 686–87). Pigment usage may also have become more complex, with many new elaborate forms of rock painting attesting to new experiments in design, motivation and meaning.

Painting also helps to ground people in both space and time. Through painting rock shelters, sheets of bark, sacred objects, bodies and bones with ochre and clay pigments, in concert with myth and ritual, Aboriginal creativity becomes united with natural creativity, intensifying Aboriginal links and bonds with the larger natural and supernatural world (Taçon 1989a, 1989b). It was always important to paint at and for ceremony, as well as the important rituals of everyone's lives. Painting and ceremony associated with one's final ritual returns the human spirit to ancestral lands. As Big Bill Neidjie concluded (Neidjie *et al* 1985: 62):

> Rock stays,
> earth stays,
> I die and put my bones in cave or earth.
> Soon my bones become earth …
> all the same.
> My spirit has gone back to my country …
> my mother.

Note

1 On one occasion I forgot to get my 4WD Toyota smoked and painted, not wishing to interrupt the middle of proceedings while I fetched it. The man who had passed away had been riding in it a few days before. The vehicle was soon plagued by flat tyres and a series of minor breakdowns, and eventually had to be carted off to Darwin for a complete overhaul.

Acknowledgments

This paper is dedicated to the late Big Bill Neidjie, a very good friend and teacher. Bill's family and other Aboriginal people from Kakadu National Park are thanked for teaching me so much about art and life since 1981. In this vein, I am grateful to the peoples of western Arnhem Land, central Arnhem Land, the Keep River region, various parts of the Kimbereley and the Waanyi of north-west Queensland for additional instruction, insight and knowledge. The Australian Museum supported my research as part of its ongoing commitment to Australian Indigenous studies. Nicole Boivin is thanked for the invitation to write a paper for this special volume on soils, stones and symbols. I am grateful to Nicole Boivin and Mary Ann Owoc and other referees for comments that improved this paper.

References

Adcock, GJ, Dennis, ES, Easteal, S, Huttley, GA, Jerniin, LS, Peacock, WJ and Thorne, A (2001) 'Mitochondrial DNA sequences in ancient Australians: implications for modern human origins' 98(2) *Proceedings of the National Academy of Sciences* 537–42

Basedow, H (1907) 'Anthropological notes on the coastal tribes of the Northern Territory', 31*Transactions of the Royal Society of South Australia* 1–62

Bowler, JM, Johnston, H, Olley, JM, Prescott, JR, Roberts, RG, Shawcross, W and Spooner, NA (2003) 'New ages for human occupation and climatic change at Lake Mungo, Australia' 421 *Nature* 837–40

Brumm, A (2000) 'The power and the story of the stone: meaningful aspects of the Aboriginal edge-ground stone hatchet', unpublished BA Honours thesis, James Cook University

Campbell, J (1834) 'Geographic memoir of Melville Island and Port Essington on the Cobourg Peninsula, Northern Australia, with some observations on the settlements which have been established on the north coast of New Holland' 4 *Journal of the Royal Geographic Society of London* 129–81

Carrington, F (1890) 'The rivers of the Northern Territory' 2 *Proceedings of the Royal Geographic Society of Australia (South Australia Branch)*

Chaloupka, G (1993) *Journey in Time: The World's Longest Continuing Art Tradition*, Chatswood, Sydney: Reed

Clark, G (1986) *Symbols of Excellence: Precious Materials as Expressions of Status*, Cambridge: CUP

David, B, McNiven, I, Attenbrow, V, Flood, J and Collins, J (1994) 'Of Lightning Brothers and White Cockatoos: dating the antiquity of signing systems in the Northern Territory, Australia' 68 *Antiquity* 241–51

Dortch, CE (1979) 'Devil's Lair, an example of prolonged cave use in southwestern Australia' 10 *World Archaeology* 258–79

Earl, W (1846) 'On the Aboriginal tribes of the northern coast of Australia' 16 *Journal of the Royal Geographic Society of London* 239–51

Elkin, AP (1930) 'The Rainbow-Serpent myth in north-west Australia' 1(3) *Oceania* 349–52

Elkin, AP (1944) *Aboriginal Men of High Degree*, Sydney: Australasian Publishing Co

Gage, J (1999) 'Did colours signify? Symbolism in the red' 9(1) *Cambridge Archaeological Journal* 110–12

Henshilwood, CS, d'Errico, F, Yates, R, Jacobs, Z, Tribolo, C, Duller, GAT, Mercier, N, Sealy, JC, Valladas, H, Watts, I and Wintle, AG (2002) 'Emergence of modern human behavior: Middle Stone Age engravings from South Africa', *Science* 10.1126/science.1067575

Horton, D (1994) 'Ochre', in Horton, D (ed), *Encyclopedia of Aboriginal Australia*, Canberra: Aboriginal Studies Press, p 820

Jones, R (ed) (1985) *Archaeological Research in Kakadu National Park*, Canberra: Australian National Parks and Wildlife Service (Special Publication 13)

McCarthey, FD (1955) 'Notes on the cave paintings of Groote and Chasm Islands' 5 *Mankind* 68–75

McKnight, CC (1969) *The Farthest Coast*, Melbourne: Melbourne University Press

Morphy, H (1991) *Ancestral Connections: Art and an Aboriginal System of Knowledge*, Chicago: University of Chicago Press

Morphy, H (1998) *Aboriginal Art*, London: Phaidon

Mulvaney, DJ (1969) *The Prehistory of Australia*, London: Thames & Hudson

Mulvaney, DJ (1978) 'Creativity in the Aboriginal past', in Edwards, R (ed), *Aboriginal Art in Australia*, Sydney: Ure Smith, pp 11–20

Mulvaney, DJ and Kamminga, J (1999) *Prehistory of Australia*, St Leonards, Sydney: Allen and Unwin

Neidjie, B (1989) *Story About Feeling*, Taylor, K (ed), Broome: Magabala Books

Neidjie, B, Davis, S and Fox, A (1985) *Kakadu Man*, NSW: Mybrood P/L

O'Connor, S (1995) 'Carpenter's Gap rockshelter 1: 40,000 years of Aboriginal occupation in the Napier Ranges, Kimberley, WA', *Australian Archaeology* 40, 58–59

O'Connor, S, Veth, P and Campbell, C (1998) 'Serpent's Glen; a Pleistocene archaeological sequence from the Western Desert' 46 *Australian Archaeology* 12–22

Robert, WCH (1979) *The Dutch Explorations, 1605–1756, of the North and Northwest Coast of Australia*, Amsterdam: Philo

Roberts, R, Jones, R and Smith, MA (1990) 'Thermoluminescence dating of a 50,000-year-old human occupation site in northern Australia' 345(6271) *Nature* 153–56

Roberts, R, Jones, R and Smith, MA (1993) 'Optical dating at Deaf Adder Gorge, Northern Territory, indicates human occupation between 53,000 and 60,000 years ago' 37 *Australian Archaeology* 58

Roberts, R, Yoshida, H, Galbraith, R, Lastett, G, Jones, R and Smith, MA (1998) 'Single-aliquot and single grain optical dating confirm thermoluminescence age estimates at Malakunanja II rock shelter in northern Australia' 16 *Ancient TL* 19–24

Rose, D (2000) [1992] *Dingo Makes Us Human*, Cambridge: CUP

Rose, D (2002) 'The dead, the missing, the lost, and the voiceless: some thoughts on extinction from a dingo perspective', paper presented at CHAGS 9, Edinburgh, at www.abdn.ac.uk/chags9/1-rose.htm

Sale, KM (1995) 'Making the old brighter: Aboriginal re-marking of rock pictures', in Ward, GK and Ward, LA (eds), *Management of Rock Imagery*, Melbourne: Occasional AURA Publication 9, pp 128–40

Smith, MA (2000) 'Ochres', in Kleinert, S and Neale, M (eds), *The Oxford Companion to Aboriginal Art and Culture*, Oxford: OUP, pp 665–66

Smith, MA, Bird, MI, Turney, CSM, Fifield, LK, Santos, GM, Hausladen, PA and di Tada, ML (2001) 'New ABOX AMS–14C ages remove dating anomolies at Puritjarra rock shelter' 53 *Australian Archaeology* 45–47

Smith, MA, Fankhauser, B and Jercer, M (1998) 'The changing provenance of red ochre at Puritjarra rock shelter, central Australia; Late Pleistocene to present' 64 *Proceedings of the Prehistoric Society* 275–92

Spencer, B and Gillen, FJ (1912) *Across Australia*, London: Macmillan

Taçon, PSC (1989a) 'Art and the essence of being: symbolic and economic aspects of fish among the peoples of western Arnhem Land, Australia', in Morphy, H (ed), *Animals into Art*, 236–50, London: Unwin Hyman

Taçon, PSC (1989b) 'From rainbow snakes to 'x-ray' fish: the nature of the recent rock painting tradition of Western Arnhem Land, Australia', unpublished PhD thesis, Australian National University, Canberra

Taçon, PSC (1991) 'The power of stone: symbolic aspects of stone use and tool development in Western Arnhem Land, Australia' 65 *Antiquity* 192–207

Taçon, PSC (1999) 'All things bright and beautiful: the role and meaning of colour in human development' 9(1) *Cambridge Archaeological Journal* 120–26

Taçon, PSC and Brockwell, S (1995) 'Arnhem Land prehistory in landscape, stone and paint' 69 *Antiquity* 676–95

Taçon, PSC, Wilson, M and Chippindale, C (1996) 'Birth of the Rainbow Serpent in Arnhem Land rock art and oral history' 31 *Archaeology in Oceania* 103–24

Taylor, L (1987) '"The same but different": social reproduction and innovation in the art of the Kunwinjku of western Arnhem Land', unpublished PhD thesis, Australian National University, Canberra

Taylor, L (1990) 'The Rainbow Serpent as visual metaphor in western Arnhem Land' 60 *Oceania* 329–44

Taylor, L (1996) *Seeing the Inside: Bark Painting in Western Arnhem Land*, Oxford: OUP

Thorne, AG (1971) 'Mungo and Kow Swamp: morphological variation in Pleistocene Australians' 8(2) *Mankind* 85–89

Thorne, AG (1975) 'Kow Swamp and Lake Mungo', unpublished PhD thesis, University of Sydney

Thorne, AG, Grün, R, Mortimer, G, Spooner, N, Simpson, J, McCulloch, M, Taylor, L and Curnoe, D (1999) 'Australia's oldest human remains: age of the Lake Mungo 3 skeleton' 36(6) *Journal of Human Evolution* 591–612

Turney, CSM, Bird, MI, Fifield, LK, Roberts, RG, Smith, MA, Dortch, CE, Grün, R, Lawson, R, Miller, MH, Dortch, J, Creswell, RG and Ayliffe, LK (2001) 'Early human occupation at Devil's Lair, southwestern Australia 50,000 years ago' 55 *Quaternary Research* 3–13

White, JP and O'Connell, JF (1982) *A Prehistory of Australia, New Guinea and Sahul*, Sydney: Academic

Wilkins, GH (1929) *Undiscovered Australia*, London: GP Putnam's Sons

– 3 –

FROM THE EARTH

MINERALS AND MEANING IN THE HOPEWELLIAN WORLD

Douglas K Charles, Julieann Van Nest and Jane E Buikstra

Given the nature of preservation in the relatively moist temperate climate of eastern North America, pre-Columbian archaeological remains exist almost entirely as manifestations of the mineral world. The lithic assemblages archaeologists find today were produced by direct percussion modifications of crystalline silica, while pottery vessels were created by thermally altering clay minerals into ceramic fabrics. Domestic architecture is preserved only as differently coloured or textured sediments filling empty post holes, the wooden posts having rotted or been removed. Dietary remains too are recoverable because of their mineral content: animal bone maintains its integrity only to the extent that its hydroxyapatite crystalline structure remains intact, as acids in the soil rapidly degrade the organic collagen substrate and will eventually dissolve the entire bone; plants are generally preserved only after they have been burned, usually accidentally, leaving a carbon and ash shell. Indeed, like animals, the people themselves exist today only as mineralised bone. In addition, the numerous mounds and earthworks studied by archaeologists were constructed from transported sediments, and carefully crafted objects of non-local materials such as copper, mica and obsidian were transported, exchanged and finally buried as grave goods. This mineralogical characterisation of the archaeological record reflects *our* Western scientific categories, however, and we might reasonably ask whether, when the Hopewell phenomenon spread across the Eastern Woodlands of North America 2,000 years ago during the Middle Woodland Period, minerals actually meant anything to the Middle Woodland 'Hopewellians'.

In this chapter we argue that it is important to differentiate between the remains associated with the Middle Woodland chronological period of the Eastern Woodlands of North America and those specifically related to what has been termed the Hopewell phenomenon that also characterises much of the same time period and geographical area. In particular, we argue that Hopewell is an archaeologically defined manifestation of social, political and economic activity, not a shared belief system or formally organised trade network. We subsequently explore aspects of the dominant material manifestations of Hopewell: earthwork construction and the circulation of non-local materials and objects. Focusing in particular on the region surrounding the lower Illinois River, we suggest that historically recorded Native American myths and beliefs help to explain the form of the earthworks and the sediments from which they were constructed. We also argue that the portable objects and substances of Hopewell were chosen for their extra-Hopewell occurrence and for their conspicuousness and/or durability. Minerals do indeed appear to have been important to Middle Woodland people, and the properties of the minerals that they used to communicate with each other across time and space also serve to communicate something of their understanding of the world to us today.

MIDDLE WOODLAND AND HOPEWELL: SETTING THE STAGE

The Woodland Period of eastern North America (**Figure 3.1**) is essentially characterised by the advent of pottery and an increasing reliance on horticulture. In general the period begins at approximately 3000 BP and ends around 1000 BP, when maize-based Mississippian chiefdoms emerged in the Midwest and Southeast, but the initial appearance of ceramics and maize varied by several centuries in different regions of the Eastern Woodlands. For example, pottery was developed first in the Southeast, and, other than the inclusion of maize in the diet, the Indians encountered by the first European colonists in New England were pursuing Late Woodland lifeways. There is evidence of horticulture in the preceding Archaic Period, but the practice intensified during the Early Woodland and was well established by the beginning of the Middle Woodland Period. The beginning and end of the Middle Woodland Period are distinguished from the Early Woodland and Late Woodland Periods, respectively, by differences in ceramic, lithic and funerary styles, but the dates and particular styles vary regionally. The dates of the Hopewell phenomenon (see below) are more or less coterminous with those of the Middle Woodland Period, in general between 2150 BP and 1550 BP, but again the dates are variable by region. For example, in the vicinity of the lower Illinois River valley, the focus of this chapter, the Middle Woodland begins about 2050 BP, the earliest Hopewell materials are dated to about 2000 BP, and both end by around 1700 BP.

Hopewell refers to a specific manifestation of material culture characterised by extensive and elaborate earthworks, partly funerary in nature, and the acquisition, transportation, exchange and frequent burial as grave goods of items of non-local origin. As opposed to the regionally heterogeneous character of Middle Woodland material culture, there is some stylistic homogeneity apparent in Hopewell earthworks and objects. The distribution of Hopewellian materials encompassed the core of the Eastern Woodlands, extending from the Appalachian Mountains in the east to the Mississippi River and beyond in the west, and from the Great Lakes in the north to the Gulf of Mexico in the south (**Figure 3.1**). The following sections will briefly explore the nature of the Hopewell phenomenon. We will also argue that we should distinguish between

Figure 3.1: Eastern United States showing the major rivers and the extent of the Hopewell phenomenon. The Archaic → Woodland → Mississippian chronological framework of the Eastern Woodlands is conventionally applied to most of the area shown on the map from Kansas eastward. The area surrounding the confluence of the Illinois, Mississippi and Missouri rivers is shown in Figure 3.2.

the Middle Woodland Period, defined largely in terms of subsistence technology, and the Hopewell practices of some Middle Woodland societies.

A brief history of Hopewell archaeology

The most famous Hopewellian earthworks are those found in what is now southern Ohio. In the 19th century, these large mounds and huge geometric enclosures, and the beautiful and elaborate artefacts they contained, were the basis for a mythical race of Mound Builders presumed to have populated the continent before the Indians (Silverberg 1968), an account still manifest in the Book of Mormon. The earthworks and artefacts were also the primary subject matter of the initial publications of the American Antiquarian Society (Atwater 1820) and the Smithsonian Institution (Squier and Davis 1848). The identity of the Mound Builders and the antiquity of people in the Americas were the two issues instrumental in the development of professional archaeology in the United States by the end of the 19th century (Meltzer 1985).

Conceptions of Hopewell continued to change over time (Van Gilder and Charles 2003). Cyrus Thomas (1894) recognised different spatial distributions of mound building cultures, while Warren K Moorehead (1899) and William C Mills (1904) introduced temporal distinctions. James B Griffin (1943: 307) noted that, in fact, Hopewell *was* the earthworks and exotic artefacts, that if 'the highly complex burial cult and the associated artifacts are taken away from the Hopewell culture little that is diagnostic or spectacular remains'. Thorne Deuel (1952) distilled this conception into the notion of Hopewell as a set of religious and political ideas, with associated material manifestations, overlying the everyday life of local Middle Woodland cultures, a construct formalised by Joseph Caldwell (1964) as the Hopewell Interaction Sphere. Writing as a founding member of the New Archaeology, Stuart Struever (1964, 1968; Struever and Houart 1972) recast the interaction sphere in economic terms. This idea has been variously developed to incorporate notions of central place and complex social organisation (eg Seeman 1979) or strategic alliance building (eg Braun and Plog 1982). While there has recently been some reconsideration of the ideological content and function of Hopewellian earthworks and artefacts (eg Buikstra *et al* 1998; Byers 1998; Hall 1997; Seeman 1995; see also Hall 1977, 1979), archaeologists have generally given explanatory priority to Middle Woodland settlement and subsistence (see eg various contributions in Anderson and Mainfort 2002; Brose and Greber 1979; Charles and Buikstra forthcoming; Dancey and Pacheco 1997; Pacheco 1996).

Rethinking Hopewell

Considering a long chronological sequence and focusing on the lower Illinois River valley, Buikstra and Charles (1999; Charles and Buikstra 2002; see also Charles 1995; Charles and Buikstra 1983) have argued that the locations of Middle Woodland/Hopewell social and ceremonial gatherings and activities should be understood in the context of similar landscape use beginning as early the Middle Archaic Period. Throughout the Middle and Late Archaic Periods (c 6000–4000 BP and 4000–3000 BP, respectively), phases of relatively high residential mobility alternated with phases of substantially reduced mobility. During the more mobile phases, groups from a wide region periodically gathered at elevated ridges and terraces in the floodplain to bury their dead and, perhaps more importantly, exchange information and materials. Significant deposits of bannerstones (atlatl weights) and projectile points occur as grave goods or in separate caches at these sites (eg Hassen and Farnsworth 1987; Titterington 1950). During the alternate periods, when mobility was curtailed, local community cemeteries were situated on the margins of the bluffs overlooking the river valley. The density of material associated with these sites is roughly one-tenth that of the floodplain sites, presumably reflecting reduced levels of circulation of these goods (Charles 1995). The Middle Woodland Period is distinctive in that both locations were used at the same time. Buikstra and Charles (1999; Charles and Buikstra 2002) argue, however, that the social patterning is maintained, that is, local communities utilised the bluff top cemeteries such as Elizabeth and Gibson/Klunk, while multiple communities gathered at the floodplain sites such as Kamp and Mound House (**Figure 3.2**). It appears that the two different locations reflect the emergence at this time of two distinct levels or networks of social interaction, one focused on establishing and maintaining the local community, the other involving intercommunity interaction and integration.

Taking the present or near present as a starting point and working backwards through time, Robert Hall (1997) makes a strong case for linking the ethnographically and ethnohistorically known Calumet ceremonialism of Plains tribes to the Hopewell phenomenon. Calumets were long decorated stems, and

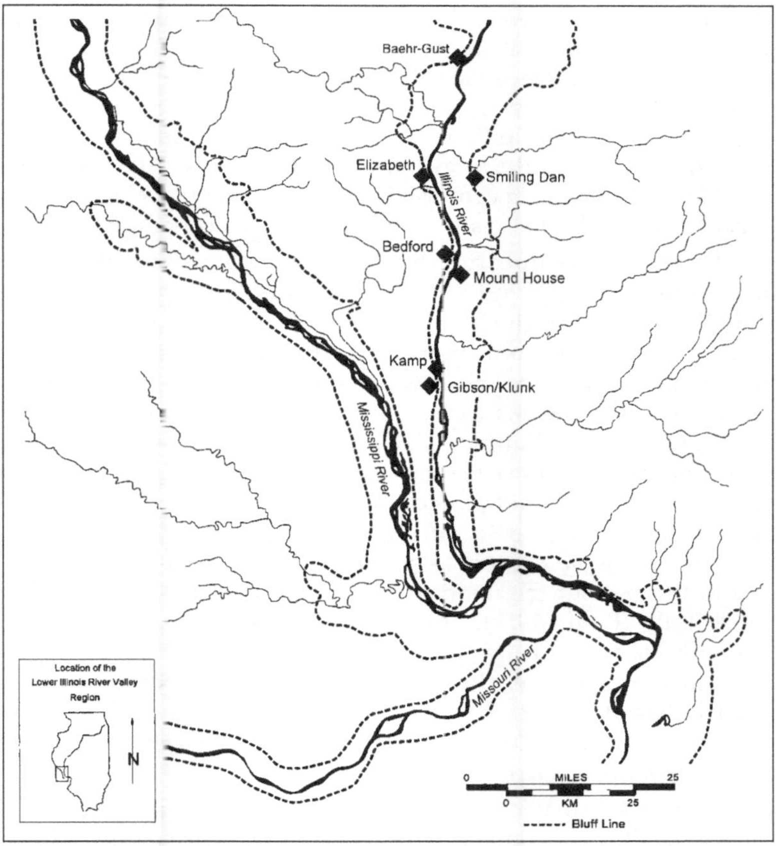

Figure 3.2: Middle Woodland sites mentioned in text: Smiling Dan is a hamlet in a small valley tributary to the Illinois; Elizabeth, Bedford and Gibson/Klunk are bluff top cemeteries; Baehr-Gust, Mound House and Kamp are floodplain ceremonial sites.

calumet pipes had a bowl for smoking attached at the end. The ceremonies were focused around the calumet and its symbolism:

> Calumet ceremonialism was a point of intersection of many aspects of Indian life. It functioned on an intertribal, political level to establish and maintain peaceful relationships between unrelated groups by creating fictions of kinship between the principals involved. It functioned in the area of economic adaptation by cushioning villages or bands against the dangers of crop failure or changes in the availability of game animals, doing so by establishing friendly patterns of exchange with neighbors upon whom one might fall back in time of need. And it functioned at the social level by providing means by which a capable and worthy person could rise in status through personal achievements. It is not difficult to imagine a similar network of relationships among Hopewell societies. (Hall 1997: 156–57)

As we shall see, these are just the kinds of functions that can be attributed to the Hopewell phenomenon, and Hall (eg 1997: fig 193) goes so far as to see direct historical connections between Woodland Period mortuary ritual and the Calumet, Sun Dance and Soul Keeping ceremonies known from Plains tribes in the Historic Period. Our conceptions of Hopewell have been greatly influenced by our Euro-American biases, such that archaeologists have tended to see Hopewell as a high point or efflorescence of culture, due to the proliferation of 'sophisticated' material culture at the time (Van Gilder and Charles 2003). Implicit in Hall's argument, in contrast, is the notion that Hopewell and Calumet ceremonialism was in the end really about politics and economics.

In Hall's terms, the question becomes: why were political alliances, subsistence buffering and individual status suddenly so important in the American midcontinent beginning approximately 2,000 years ago? These were surely not new concerns, and mortuary ritual and exchange of non-local valuables seem to have been the media for dealing with them since at least the Middle Archaic four millennia earlier (Brown 1985; Buikstra and Charles 1999; Charles 1995; Charles and Buikstra 2002). From this perspective, the appearance of Hopewell indicates not a rise in social complexity *per se*, but rather the onset of a period of marked social and political activity (Van Gilder and Charles 2003). The impetus for this rise in activity was arguably the demographic transformation of the landscape, as populations began to aggregate in larger river valleys throughout the midcontinent at the end of the Early Woodland Period, c 2050 BP (Braun 1987). This population redistribution was related to improvements in horticultural practices, in particular the development of ceramic technology. The river floodplains were the ideal locations for this new horticulture based on starchy seeds such as sumpweed (*Iva annua*), chenopodium (*Chenopodium berlandieri*), erect knotweed (*Polygonum erectum*), and maygrass (*Phalaris caroliniana*).

The demographic changes were particularly dramatic in the lower Illinois River valley. The Illinois floodplain evolved over the course of the Holocene, but it had effectively attained its current horticulture-conducive landform equilibrium and flooding regime by this time (Hajic 1990; Van Nest 1997). Apparently the region had been largely abandoned by the end of the Early

Woodland Period (Charles *et al* 1986; Farnsworth and Asch 1986), setting the stage for a marked demographic transformation. The earliest Middle Woodland occupation occurred in the northern end of the valley, with infilling proceeding southward over the course of the next 300 years (Charles 1992; Kut and Buikstra 1998). Population densities were uniformly dense throughout the succeeding Late Woodland Period (Charles 1992).

Under these subsistence and demographic conditions, the political and territorial relations between new migrants and earlier pioneers would have had to have been constantly negotiated and renegotiated (Charles 1992, 1995; Charles and Buikstra 2002). The decreasing mobility resulting from populations concentrating in the valley would have increased tensions among communities, intensifying the need for dispute mediation and alliance formation. Reduced mobility would also have increased subsistence risk due to seasonal or annual fluctuations in resource availability, since movement to more productive areas would have been increasingly precluded. Establishing and maintaining relations with neighbours through the exchange of Hopewell paraphernalia would have facilitated the development of avenues for negotiations and for buffering risk. In general, the acquisition, by travel or trade, and subsequent exchange of Hopewell items and raw materials may be seen as part of a strategy through which individuals (and by extension the lineages and communities of which they were members) established and maintained prestige and negotiated alliances and other reciprocal relationships. The construction and use of earthworks served as the venue for the exchange and disposal (as grave goods) of Hopewell material. Thus, Hopewell is 'about' the increased intensity of negotiation and contestation of social, political and economic relations among individuals, residential groups and wider communities.

The end of Hopewell in the lower Illinois River valley corresponds to the slowing down and eventual cessation of this demographic transformation of the landscape. By around 1700 BP the valley had become crowded, and occupation was beginning to expand into the uplands and tributary valleys (Charles 1992, 1995; Charles and Buikstra 2002). Biological distance measures suggest that kinship networks had stabilised geographically and had begun to expand by this time (Conner 1990). Simply put, the previous Middle Woodland level of negotiation among individuals and communities, utilising the media of Hopewell, was no longer necessary. Social, political and economic interaction in a more stable demographic landscape was now being conducted along kinship webs visible only through biological, rather than archaeological, analyses.

Middle Woodland vs Hopewell material culture

Nevertheless, archaeologists have tended to distinguish between the physical remains of Middle Woodland *Period* subsistence and settlement and those of the Hopewell *Interaction Sphere*, privileging the former's economic and adaptive importance. The Middle Woodland Period is defined by local ceramic and lithic styles that differed regionally throughout the Eastern Woodlands of North America. Although associated with extensive horticulture in some areas, such as westcentral Illinois, the contribution of gardening to the gathering and hunting subsistence economy also varied from region to region. Similarly, the extent to which local populations participated in the Hopewell Interaction Sphere was highly variable. The Middle Woodland Period remains of subsistence and settlement share several characteristics: (1) they tend to have been of local origin; (2) they generally consist of materials that were abandoned (eg refuse) or dismantled (eg structures); and (3) they were either acquired because of their specific technological properties (eg chert and clay) or incidentally (ie contained in plants and animals), or they are simply taphonomic byproducts (eg postmolds). Conversely, many of the materials we attribute to the Hopewell Interaction Sphere – for example, the sediments used in earthwork construction and the raw materials fashioned into the styles characteristic of Hopewell – share different characteristics: (1) they were transported considerable distances, or, especially in the case of earthwork fills, did not always come from the immediate vicinity; (2) they were often intentionally deposited (*sensu* Schiffer 1972, 1987) in grave caches or in a patterned manner; and (3) they were seemingly chosen for visibility or durability, or for symbolic reasons. Thus, whatever meanings (theirs or ours) we may ascribe to these archaeological remains, and whatever the significance of those meanings, the spatial and contextual dimensions of the acquisition and deposition of those materials we recognise as Middle Woodland differed from a significant subset of those we recognise as Hopewell. Similarly, the remains differ in kind, with those of Middle Woodland typically relating to the acquisition and processing of resources, and those of Hopewell relating to long term and/or long distance communication activities. These characterisations are not meant to assign differing values to Middle Woodland Period subsistence practices or Hopewell Interaction Sphere materials. Rather, we suggest that the properties of the minerals we associate with Middle Woodland and Hopewell differ in ways that relate to the manner in which they were implicated in daily life. That local Middle Woodland materials and Hopewell styles were at times more complexly integrated in Illinois than this model would suggest is, however, clear from the recent source identification of local ceramic clays and nearby flint clay outcrops (from which smoking pipes were carved), the products of which were decorated in pan-Hopewellian motifs (Fie 2000; Hughes *et al* 1998). Nonetheless, the contrasts we have defined provide a perspective from which to explore Middle Woodland life and the nature of the Hopewell phenomenon. In the discussion that follows, we will examine earthworks first, followed by exchanged materials.

EARTHWORKS: STRUCTURE, COMPOSITION AND MEANING

The main distribution of Hopewellian earthworks forms an arc running from north-east to south-west, largely following the Ohio and lower Mississippi rivers (**Figure 3.1**). In Ohio, the earthworks consist of huge mounds (eg Greber 1979), as well as circular, square, octagonal and conjoined geometric enclosures (eg Lepper 1996), and embankments that delineate hilltops or other remote landforms (eg Connolly 1998; Riordan 1998). Two monstrous mounds and some small embankments are known from the Ohio River valley in southern Indiana

(Ruby 1997; Seeman 1995). The lower Illinois River valley (**Figure 3.2**) is lined with hundreds of burial mounds (Charles 1992), and there is a probable Hopewell ovoid or elliptical earthwork near to where the Illinois joins the Mississippi River (Van Nest forthcoming). Along the lower Mississippi River, a number of platform mounds and semicircular enclosures have been recorded and excavated (eg Mainfort 1996). Numerous burial mounds are found on either side of this riverine corridor, from Michigan, Wisconsin, Iowa and Kansas to the north and west to Alabama and Georgia in the south-east.

These earthen structures occur in a variety of forms. The geometric earthworks of Ohio, for example, are highly diverse, though seemingly variations on a theme (Squier and Davis 1848). The mounds, which are ubiquitous throughout the Hopewellian world, are similar only in their final form: a round or oval dome of earth. They range in diameter from 2 to 100 m, and their heights vary from a few centimetres to over 10 m (Charles 1992), with even larger examples in Ohio and elsewhere (eg Greber *et al* 1983; Seeman 1995). The internal structure of mounds seems almost infinitely variable. In Ohio large mounds covered dismantled wooden charnel structures, while in Illinois mounds enclosed smaller, earthen crypts (Brown 1979). In Tennessee rectangular platform mounds were constructed in six or more stages (Mainfort 1988), while in the Illinois valley, there are differences between mounds found in the central and lower valleys, and between those situated on floodplain and bluff top locations (Bullington 1988).

Dozens of earthen mounds have been excavated along the lower Illinois River valley since the 19th century (eg Buikstra *et al* 1998; Charles *et al* 1988; Farnsworth and Wiant forthcoming; Fowke 1905; Henderson 1884; Perino 1968, 1973; Struever 1960; Titterington 1935), the vast majority of them dating to the Middle or Late Woodland Periods (c 2000–1700 BP and 1700–1000 BP, respectively, in this region). There is also a wealth of relevant environmental, geological, subsistence and settlement information that has been collected for this region, mainly since the 1970s (eg Asch and Asch 1985; Brown and Vierra 1983; Butzer 1977; Hajic 1985, 1990; Hewitt 1983; Neusius 1986; Rubey 1952; Stafford and Sant 1985; Wiant and McGimsey 1986; Styles 1981; Styles 1985; Van Nest 1997). From the excavations, a consensus picture of Illinois valley Hopewell mounds has coalesced (Braun 1979; Brown 1979; Bullington 1988; Buikstra 1984: fig 9.2; Buikstra and Charles 1999: fig 9.6; Buikstra *et al* 1998; Charles 1992: fig 1; Perino 1968; Tainter 1977; but see Van Nest forthcoming for an account of the range of variability). The illustration (**Figure 3.3**) and description of a 'typical' bluff top mound given here, it should be noted, does not correspond to any actual mound that has been excavated; it is rather the modal point for a series of idiosyncratic instances. The various attributes described in what follows are not present in all cases (eg, the initial sand or loess layer may or may not be present; the crypt may or may not have an excavated subsurface portion; some crypts have a log outline one or two logs high, some do not).

Mound structure

In the area where a mound was to be constructed, the vegetation and sometimes the A-horizon were stripped from the land surface. On this exposed subsoil, a thin (2–5 cm thick) circular layer of light-coloured sand or loess was deposited. In the centre of this disk, a rectangular crypt approximately 2 m or more in length and somewhat narrower in width was constructed, either on the light sediment or over a subfloor pit of the same size and shape. The walls of the crypt were built from stacked sections of sod or turf, extending outward in what is generally referred to as a ramp. The ramp was highest in the centre around the crypt, where a removable log roof covering rested on top of the crypt. The ramp then sloped downward and outward to the land surface, ending at the edge of the circular sediment layer. The structure in this configuration was the operational mortuary facility (**Figure 3.3**). Some of the deceased were placed in the crypt for a period of time, during which decomposition of the soft tissue would take place. The skeletal elements were then removed, wrapped in a small bundle and deposited in a pit dug into the surface of the ramp. Other dead were buried in pits or small crypts outside the ramp, but aligned tangentially to it. At a certain point in time, burial activity ceased, and the crypt and ramp structure and the surrounding graves were sealed by a mantle of earth.

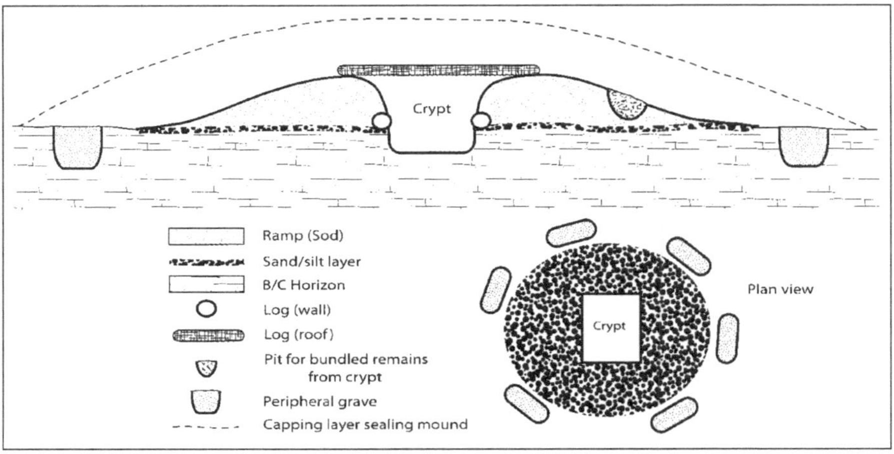

Figure 3.3: Generalised components and configuration of a lower Illinois River valley bluff top burial mound. The structures are highly variable, and not all components are always present; there may be multiple, sequential crypts, etc. Floodplain mounds follow a similar pattern, but are generally larger, usually contain multiple crypts, lack peripheral graves and ramp bundle burials, and consequently contain far smaller numbers of interments.

Sediment composition of the mounds

A closer look at the origin of sediments comprising the mounds provides suggestive evidence of symbolism inherent in the construction of the facilities. In the lower Illinois River valley, Hopewell mounds occur in one of two physiographic settings (see above; Buikstra and Charles 1999). Small community burial mounds (**Figure 3.4**) were generally situated on ridges at the margins of the bluffs overlooking the valley floor (Charles 1992). Larger mounds (**Figure 3.5**), with fewer burials, were associated with floodplain 'gathering' sites periodically occupied by multiple communities (Charles 1995). The mounds in general, and floodplain mounds in particular, are more properly seen as platforms or stages for ceremonial performances into which some burial crypts are incorporated, rather than as cemeteries *per se* (Buikstra *et al* 1998; Charles and Buikstra 2002; see also Brown 1979).

Figure 3.4: Mound 6 at the bluff top Elizabeth site. At bluff top sites, the mounds were generally arrayed along the crests of the erosional ridges formed at the margins of the bluffs, some 75–100 m above the floodplain (courtesy Center for American Archeology).

Figure 3.5: Tree-covered Mound 1 at the floodplain Mound House site, situated on a large sand ridge. In the foreground is the flat 'plaza' between the mounds. To the right of the mound is the edge of the ridge dropping into the prehistoric lake bed (which would have extended beyond the row of trees). In the distant background the bluffs rise above the floodplain (courtesy Center for American Archeology).

Initial sand/silt layers

At the bluff top Elizabeth site (**Figures 3.2, 3.4**), for instance, two mounds (Mounds 6 and 7) contained basal light-coloured layers of sediment (**Figure 3.6**) presumably derived from the Pleistocene loess that mantles the limestone bluffs on either side of the river valley (Leigh *et al* 1988). Although a source of loess existed adjacent to the mounds in the ridge on which the structures rested, the calcareous loess that was used had to have been mined from probably a metre or more below the land surface. Only one possible borrow pit (Feature 1, adjacent to Mound 4) was encountered during the substantial off-mound testing that was conducted during excavation of the site (Charles *et al* 1988), and it was not large enough to account for the amount of earth involved. Alternatively, unweathered calcareous loess may have been found closer to the surface at some distance down the side of the ridge where the slope was steeper and weathering was not as deep.

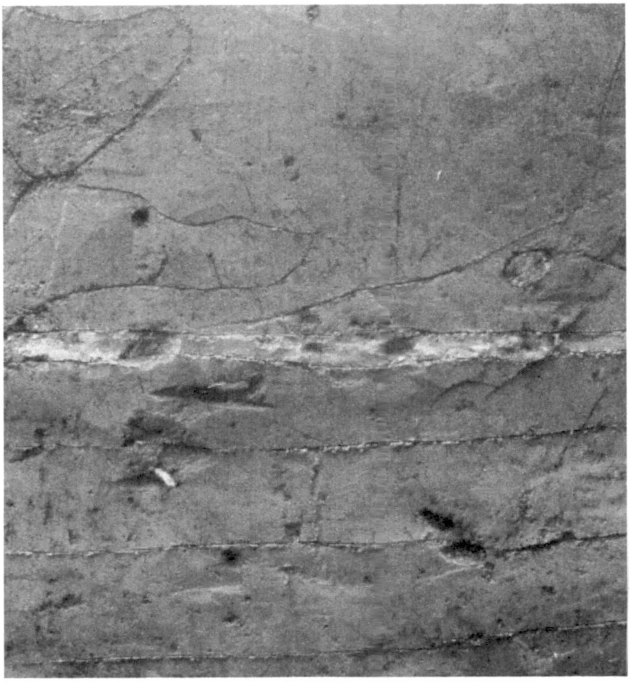

Figure 3.6: Light-coloured loess layer underlying the ramp of Mound 6 at the Elizabeth site. Moundfill (above loess layer) is weathered and loaded fills are only faintly evident. Vegetation and very upper portions of the A-horizon were stripped from the ridge surface (beneath loess layer), but remainder of A-horizon grading into B-horizon are evident (courtesy Center for American Archeology).

At the Mound House site (**Figures 3.2, 3.5**), located on a sand ridge in the Illinois floodplain, fairly pure sand was used in a similar layer under Mound 1 (Buikstra *et al* 1998). Again, the sand could have been procured in the vicinity of the mound, but only by extracting it from the Pleistocene deposits a metre or more below the surface. At the similarly situated Baehr-Gust site (**Figure 3.2**), the closest source for the sand layer was the Pleistocene terrace on the opposite side of the river, although another possibility for sand procurement would have been recently formed flood deposits (Van Nest forthcoming). Whether the colour or the source of these materials, if either, was important is not clear. The juxtaposition of light and dark sediments or other materials is widespread across the region during this time in mounds (Buikstra *et al* 1998) and in other contexts, such as artefact caches (Morrow 1991), but no potential meaning relating to this patterning is immediately obvious.

The manipulation of light-coloured sand at Mound House was, however, more complex than the simple spreading of a thin layer, as tended to occur in the bluff top mounds. The 'pristine' sand may also have been important for renewing the ceremonial surface (Buikstra *et al* 1998; Van Nest forthcoming). In the construction and utilisation sequence of Mound 1, there was a base layer of sand, but sand or silt had apparently been deposited and removed several times before this final layer was installed. In the ridge beneath the sand layer, and extending beyond its limits, were several oval rings of postmolds filled with sand or other light-coloured sediments (**Figure 3.7**). Several of the postmolds contained multiple sediments, that is different sands and silts, deposited in a manner that suggests that the post holes had been filled, re-excavated, and filled again, some in several such cycles (eg Buikstra *et al* 1998: fig 5.19). The wide spacing and shallow depths of the postmolds suggest that they were used to demarcate a space rather than to support a substantial roofed or walled structure. It appears that an array of posts was erected, then removed, and the area covered by a sand or silt layer. At a later date, the sand/silt layer was cleared away, exposing the postmolds, now filled with the light sediment set in the very dark soil of the ridge. The cycle was then repeated, with posts erected in new holes or reusing earlier ones. At least two episodes of construction involving smaller posts and without the light-coloured sediments preceded the

Figure 3.7: Sand-filled postmold beneath Mound 1 at the Mound House site. Differently textured fill (containing a higher percentage of silt) lines the base of the postmold, suggesting a previous fill episode (courtesy Center for American Archeology).

sand/silt cycles. The final layer of sand (smaller in diameter than some of the earlier layers) was the first stratum of the mound. Single episodes of post construction have been found beneath other Middle Woodland mounds in Illinois (eg Cole and Deuel 1937; Hesselberth 1945; Staab 1984; Wray and MacNeish 1961) and Ohio (eg Baby 1971; Brown 1979; Greber 1979), but we are aware of no other instances of repeated construction such as occurred at Mound House. At the Pinson site in Tennessee, however, Mound 5 'was constructed in at least six stages, with the summit of each stage covered with pale yellow sand' (Mainfort 1996: 384; see also Mainfort 1988).

Sod ramps

We have recently demonstrated that the ramps surrounding the burial mound crypts in the Illinois valley were constructed from sections of sod or turf rather than loose fill (Van Nest forthcoming; Van Nest *et al* 2001). Since the 19th century, discrete masses of earth, often of different colours and presumably deposited individually, have been characterised as loaded fills. More specifically, they have often been referred to as 'basket loading', on the assumption that each component represents an individual basketful of soil. In 1996, a clear exposure of deeply buried, and therefore well preserved, loading led to field identification of the loads as inverted blocks of sod with intact A-horizons (**Figure 3.8**). Subsequent laboratory analyses confirmed the identification (Buikstra *et al* 1997; Smith *et al* 1997; Van Nest *et al* 2001). It is clear from examination of published photographs that much, perhaps most, of the 'basket loading' reported in other earthworks in the region, for example in Klunk Mounds 1 and 2 (**Figure 3.3**; Perino 1968: figs 4b, 20) and Elizabeth Mound 7 (Leigh *et al* 1988: fig 5.26), are in reality stacked sod blocks.

Figure 3.8: Edge of a ramp in Mound 1 at the Mound House site showing individual pieces of inverted sod.

More labour was involved in the use of sod as a construction material than is immediately apparent: not just any sod would do. The following arguments are developed in detail elsewhere (Van Nest *et al* 2001; see also Van Nest forthcoming). In order to ascertain the origin of mound sods, soil samples were excavated or cored from mounds at three sites. All three mound groups are located in the floodplain, Baehr-Gust and Kamp on alluvial fans, and Mound House on a sand ridge (**Figure 3.2**). In addition, cores were taken from potential source locations. At all three sites, the sod blocks were cut from grassy areas, but the sites themselves would have been forested when mound construction was initiated. At Baehr-Gust and Kamp, the sod had developed on the loessic soils that cover the uplands and higher terraces, not on alluvial fan sediments like those beneath the mounds. At Mound House, the loam and silt loam sod blocks contrast with the sandy loam and loamy sand deposits of the ridge. The mixed, unsorted nature of the sod substrate in the latter case suggests an origin in flood deposits that would have been found in close proximity to the river.

Not only are the sods comprised of allogenetic parent materials, that is, materials transported from some location other than the immediate surroundings of the mound, but they are also primarily anthropogenic in origin. The A-horizons of the sod blocks are dark coloured (very dark grayish brown to black) but thin (3–10 cm in thickness). The lack of any significant development of sub-A soil horizons suggests that the soils represented in the sods are very young, and could conceivably have formed within a decade or two. Grass dominates the phytolith assemblages of the sod A-horizons, a finding compatible with recovered pollen assemblages that suggest an open or disturbed habitat. Moreover, the soils represented by the sods contrast markedly with those documented in both below- and away-from-mound contexts, soils that show that these sites were forested prior to and after their use as Middle Woodland ceremonial spaces. Given ecological conditions for this part of Illinois, the most likely scenario that enabled the formation of sods like these involves opening up the forest canopy sufficiently to allow the development of a grassy cover. Whether girdling, fires, or a combination of both initially accomplished this, it almost certainly entailed subsequent anthropogenic maintenance by burning, because neither primary nor secondary succession in these environments would lead naturally to a grassy soil cover. Also significant is the uniformly thin A-horizon of the sods despite development into very different parent materials at different sites (for example, loessial silt loams at Baehr and Kamp, loams and silt loams at Mound House). All of these facts suggest that the sods were obtained from regularly maintained open spaces and raise the question of whether the spaces themselves were important.

Mounds and the Hopewellian cosmos

Two widespread cosmological conceptions seem to provide insight into meanings that may have been attributed to these mounds and their components. The first is the fairly uniform picture of the cosmos that characterised the belief systems of Native Americans who, at the time of colonisation, lived in the south-eastern region of what is now the United States, an area that also made up a large portion of the Hopewellian world:

> The Southeastern Indians conceived of This World as a great, flat [circular] island resting ... on the surface of the waters ... The Southeastern cosmos consisted of three worlds: in addition to This World, an Upper World existed above the sky vault, and an Under World existed beneath the earth and the waters. (Hudson 1976: 122)

This tripartite vision of the world appears to be widespread and deeply rooted in the Native American past, extending even beyond their history in the Americas. Similar constructs are found, for example, among Plains Cheyenne and across northern Siberia (Schlesier 1987). The structure of Hopewell burial mounds in Illinois seems to mirror this conception of the universe (Buikstra and Charles 1999): it may be argued that the Upper World existed above the burial mound, while the removal of the vegetation or A-horizon prior to mound construction exposed the Under World; the layer of light-coloured sediment would thus represent the circular island of This World.

A second common belief, the myth of the Earth Diver, relates to sod (Hall 1979; Leeming and Page 1998). The Earth Diver story is similarly not confined to Native Americans, but is seemingly distributed across Europe, Asia and North America, from Finland to the south-eastern United States (Count 1952):

> The basic theme is that of an animal that dives into the depths of the sea to retrieve mud that then expands to create the land. There may be attempts by several animals, but one finally succeeds with the mud clinging to his little paw or foot. In Asia the earth-diving animal is typically a bird, but an earth-diving bird is featured in North America only in the northern Plains. The Cherokee Earth Diver was a water beetle; a muskrat for the Ojibwas, Ottawas, Foxes, and Onondagas; a crawfish for the Shawnees and Osages; a duck for the Arikaras; a mudhen for the Cheyennes, and a red-headed duck and a turtle together for the Arapahoes. (Hall 1997: 19)

Sod, possibly referring to the Earth Diver story, still figures in the ceremonies of Plains tribes. For example, Harrod (1987: 130) suggests that in the sweat lodge rituals forming part of the Blackfoot Sun Dance, 'earth dug out from the floor of the sweat house to make a place for the heated stones was placed on one side of the structure and symbolized the earth brought up from the primal waters by the animals' (although his source, McClintock [1999: 285–87] makes the connection of sod to digging animals, such as badgers, not to the Earth Diver myth). The term 'Sun Dance' is actually a construction of travellers, historians and ethnographers that describes a number of midsummer ceremonies, ranging from mourning rites to world renewal ceremonies, performed by Plains Indians (Hall 1997: 20). These include the Arapaho Ceremony of the Offerings Lodge in which two sods representing the mud of the Earth Diver myth were placed on either side of a buffalo skull (Hall 1997: 19), and the Cheyenne New Life Lodge ceremony where five pieces of sod were arrayed around the skull (Hall 1997: fig 3.2).

Building a mound::re-enacting creation

Assuming that the sod in the ramps of Hopewellian burial mounds in the Illinois valley refers to an earlier version of the Earth Diver myth, the layer of light-coloured sand or silt at the base of the mounds may be more properly seen as a general reference to This World, a sort of plane of existence. The sod ramp then represents the actual world, the physical earth, spread out across that plane. While small slabs of sod in the Sun Dance serve to invoke the myth, the metaphorical recreation of the world on the scale of a mound entails substantially more sod. Within this framework, the activities associated with the construction and use of the mound take on even more significance.

The ramp and crypt complex did not just *represent* the cosmos: the very act of building it out of sod *re-enacted* cosmological creation. With this recognition, the process of burial also takes on new meaning. In one possible burial trajectory, the body of the deceased traverses This World as it is carried across the ramp to the crypt, which lies at the centre of the earth. There the soft (wet) tissues of the body decompose and are absorbed back into the Under World. Finally, the dry bones are gathered up and interred in the metaphorical earth, represented by the sod ramp. If and how the soft tissue/bone and/or wet/dry oppositions were meaningful to the Hopewellians are difficult questions to address, but it is likely that the distinctions were recognised and imbued with significance. The relationships of the corpse and the soul with processes of decomposition and transition in secondary burial are widely documented ethnographically and were theorised almost a century ago by Hertz (1960; see also Metcalf and Huntington 1991). Secondary burial is known from Archaic cemeteries in the region dating back to at least 4000 BP (Charles and Buikstra 1983). The frequency of disarticulated, rearticulated and bundled skeletal remains, and the prominence of crypt, tomb and charnel facilities in Woodland and subsequent Mississippian Period funerary practices (eg Brown 1971, 1979; Fowler 1978), suggest that secondary burial was a particularly important component of mortuary ritual for over a millennium, beginning at least as early as the Middle Woodland era.

A second burial trajectory involving interment in peripheral graves (**Figure 3.3**) has also been observed in many of these mounds. These two trajectories appear to represent separate lineages within the community, with the crypt track used by the dominant lineage (Brown 1981; see also Buikstra 1976; Charles 1992). The bodies in the second track, the deceased of subordinate lineages, were interred in graves that were in effect outside the world. The cosmic metaphor of the mound thus served to reinforce hierarchical social relations.

Well-developed natural sod, but with a much thicker A-horizon, would have been available in the vicinities of the mounds, but, as indicated earlier, was not selected for use. As noted, the spaces where the sod was produced may have been significant. The *creation* of sod may also have been meaningful, and involved more than just intentional burning. Some of the sod blocks at the Mound House site had been grown on soils to which a thin layer of clay-rich alluvial mud had been added. We do not know whether or not these additions were simply incidental drapes of flood mud, but we would like to suggest the possibility that clay-rich soil materials were added to the surfaces of these sand-rich soils to enhance fertility. In addition to the Hopewell phenomenon, the Middle Woodland Period witnessed a boom in horticulture. It is thus possible that the increasingly sophisticated horticultural expertise of the Hopewell people was brought to bear on the symbolic construction of mortuary facilities.

Finally, a link between fertility and mound symbolism relating to world renewal may also have been important. Robert L Hall (1979: 260) noted a quarter of a century ago that a 'strong argument can be made that some Hopewellian and other Woodland mortuary ceremonialism was in fact creation drama reenacting mythical origins'. He detailed a number of other possible connections between the Earth Diver and world renewal ceremonies, known ethnographically, and elements of Hopewell funerary ritual, including the use of black muck, marls and clays, and the importance of the cardinal directions and colour symbolism.

MINERALS FROM AFAR

In addition to the earthworks, the other spectacular material manifestation of Hopewell involves the distribution throughout the Hopewellian world of exotic raw materials and items made from them. The majority of finished (and whole) objects have been recovered from burial or cache contexts (**Figures 3.9, 3.10**; see also Brose 1985), but broken or small objects and scraps of raw material are frequently encountered in domestic sites. In earlier conceptions of the Hopewell Interaction Sphere, these objects were regarded as religious paraphernalia (eg Caldwell 1964) or as media of ritualised trade that established partnerships that were also valuable in terms of alliance building and/or subsistence buffering (eg

Figure 3.9: Hopewell copper plaque with bird imagery from the Bedford mound site (photographed by KB Farnsworth through the courtesy of the Gilcrease Institute of American History and Art, Tulsa, Oklahoma).

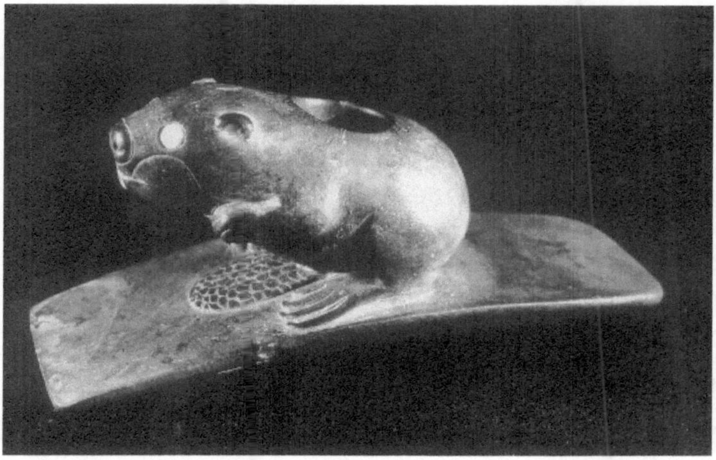

Figure 3.10: Hopewell effigy pipe from the Bedford mound site, manufactured from pipestone from northern Illinois, inlaid with pearl eyes and carved bone teeth (photographed by KB Farnsworth through the courtesy of the Gilcrease Institute of American History and Art, Tulsa, Oklahoma).

Braun and Plog 1982; Brose 1979; Struever and Houart 1972). Seeman (1995), following from the ideas of Helms (1988), has more recently argued that Hopewell materials and objects served to facilitate communication across linguistic boundaries. He also argues that the broad distribution of Hopewell styles suggests that it is more than a set of socioeconomic relations: that 'Hopewell also must be seen as an ideological system' (Seeman 1995: 122).

Many of the surviving materials specifically related to the Hopewell phenomenon consist of mineral forms obtained from the physical world: obsidian, copper, mica, galena, colourful and fine-grained cherts, and pipestone, for example. Other items, such as marine shells, riverine pearls and bear canines, might be considered to be of organic origin. However, shell and pearls are primarily calcium carbonate and are deposited externally to the soft body. Even teeth blur the boundary between the organic and the inorganic: they are the most mineralised tissues in the mammalian body, unlike bone they are not remodelled, and they protrude from the soft tissue. Thus, even within our scientific categories, shell, pearls and teeth are ambiguous and can readily be labelled as mineral rather than animal. The durability of these various materials facilitates their preservation and archaeological recovery, so our sample of Hopewell substances may be biased. On the other hand, that same durability would serve well for items that needed to travel great distances, pass through many hands, and last for generations.

Design motifs applied to ceramics, carved stone pipes and copper and mica cutouts include geometric forms, birds and other animals, and hands or claws, but it is at present difficult to make much of the meanings presumably associated with these symbols (but see DeBoer 1997; Hall 1997). Some materials, such as galena, certain cherts, pipestone and pearls, originate from specific restricted locations within the geographic domain of Hopewellian cultures. Ceramic clays in particular are more localised in terms of their sources and become 'Hopewellised' by virtue of their decorative treatment. Obsidian, copper, mica and marine shell, in contrast, come from the margins of, or even beyond, what is generally considered the Hopewell domain. In the belief systems of Native Americans of both the Southeast and the Plains, cardinal directions are often associated with colours (Hall 1997; Hudson 1976). It is tempting to see black obsidian from Wyoming in the west, red copper from the Upper Peninsula of Michigan in the north, and white marine shell from the Gulf of Mexico to the south as symbolically associated with the directions of their sources. Mica from the Appalachians in the east, the direction of the rising sun, might have been important for its reflectivity of light, rather than its colour *per se* (Helms 1995).

Things that come from elsewhere are often imbued with power or magic, of either good or evil nature (Helms 1988). Such objects can be obtained through reciprocal exchange networks, which are generally seen to characterise the Hopewell Interaction Sphere, and it is likely that such networks, which are venues for the acquisition of prestige or status, were part of Hopewell practices. Exotic items can also be obtained by travelling to their source, however, and there are important reasons for people to travel, as outlined in Helms' cross-cultural study, *Ulysses' Sail*:

> Even a cursory overview of travel motives as they are expressed in ethnographic accounts suggest that (1) reasons for travel outside of the home society are diverse, even for single cases; (2) the majority of motives expressed contain a high degree of self-realization or personal benefit; (3) many, perhaps most, expressed reasons for travel are of a *political, ideological,* or *intellectual* nature; and (4) economic benefit, when cited in the literature, frequently though not inevitably, is part of a more diverse set of motives or goals. (Helms 1988: 67; emphasis in original)

Thus, Hopewell travellers may have been shamans, traders and/or young men on a walkabout. For any of these travellers, Hopewellian objects could have served both as a medium of communication (Seeman 1995) and as the reason for the journey. They would also have served as proof of the journey, or at least of contact with distant lands. The acquisition of esoteric knowledge is often the object of travel or exchange (Helms 1988). The items themselves may or may not have had magical powers, but they certainly brought their bearers into contact with the wider world, and therefore may have related to the procurement of esoteric and powerful knowledge. In any case, whether objects were obtained through exchange or by acquisition at the source, one property that many of the Hopewellian materials share is that they were foreign, and did not come from the everyday world. The meanings of the Hopewellian objects apparently extended beyond the exoticness of the material used to make them, however. For example, local materials of similar colour were sometimes substituted for exotic minerals (eg Morrow 1991; see also Buikstra *et al* 1998). Identifiable objects, most notably bear canines, were sometimes replicated using other media, such as bone, suggesting the importance of metaphorical or metonymical associations.

MINERALS AND MEANING

The Hopewell phenomenon has long been regarded as the first great efflorescence of Native American cultures in eastern North America, but we must be careful in equating the concept of material wealth with efflorescence. We must be equally wary of overemphasising the sacred or ritual contexts of final deposition and thereby reifying Hopewell in those terms. There was indeed an increase in the acquisition and circulation of foreign and rare raw materials and in the manufacture of iconographic and otherwise decorated products from these materials as part of the Hopewell phenomenon. There was undoubtedly deep religious and magical significance surrounding these items, particularly in the funerary contexts in which we generally find them. Why the specific materials were chosen, and why the particular forms were produced, will perhaps eventually be understood in terms of the contexts and histories of specific beliefs. Nevertheless, the frequency and elaboration of these materials and earthworks at this particular time was the product of an increased intensity of social, political and economic activity corresponding to a massive redistribution of people across the Middle Woodland landscape. We tend to see the material culture of Hopewell as a manifestation of ritual practice, but that is because we are seeing sealed crypt and ramp complexes with associated human remains as cemeteries rather than as the active structures they were during most of their use life. We see obsidian, copper, mica and the other materials of Hopewell as grave goods, when they may have spent generations in circulation. In both cases we are seeing only the end points of much longer-term practices.

Hopewell and Middle Woodland do not represent oppositions of the sacred and the profane, of ritual and everyday. Meaning and ritual would have infused all aspects of Middle Woodland life. Hopewell was rather about communication across time and space, in the context of individual and group identity and status negotiation, and the properties that made Hopewell materials effective tools of communication 2,000 years ago – durability, foreignness or rarity, colour, and design or decorative plasticity – are the same properties that make them conspicuous today in the archaeological record. It should therefore come as no surprise that we can now differentiate between the products and byproducts of Middle Woodland Period settlement and subsistence and the Hopewellian mineral forms that were intentionally manipulated in social, political and economic arenas. The archaeological manifestation of Middle Woodland is largely fallout from everyday, local tasks – broken or worn out tools manufactured from the ubiquitous Burlington chert (and debitage from their production), shattered pots made from local clays, dietary refuse consisting of animal bones hardened by calcium and the remains of plants that had absorbed carbon from the atmosphere. The manipulation and transformation of these materials flowed from considerations of cultural proscriptions and prescriptions, technical efficiency, individual agency, and personal idiosyncrasies. Meaning was undoubtedly derived from and invested in these minerals, but those meanings are more difficult for us to access with confidence in prehistoric contexts.

In contrast, many Hopewellian substances are characterised by material properties like conspicuousness and durability that through a congruence of phenomenological experience, theirs and ours, begin to open up the meanings associated with these minerals. These substances include, for example, the specific soils transported to build representations of the world from which to orchestrate ceremonies and in which ultimately to house the dead, and the beautiful, unusual and often foreign glass, metal, stone and shell items that were given and received as individuals and communities negotiated their place, standing and identity. The evident intentionality inherent in the acquisition, use and disposal of these substances constitute these acts as 'events ... [that] provide us with our clearest entries into a past as experienced by its participants' (Charles and Buikstra 2002: 16). By employing an historical framework, and with the aid of ethnographic and ethnohistoric accounts, we are able to construct a plausible, perhaps even probable, account of the Hopewell phenomenon as the residue of the intense social, political and economic activity that characterised so much of Middle Woodland life during a period of great demographic and social change. The materiality of Hopewell entailed extraordinary procurement costs of acquisition and manipulation. The sods and soils were seemingly invested with cosmological meaning, and the obsidian, mica, copper and other substances may have invoked the mystery and power of distant places and secret knowledge. As such, these Hopewellian minerals would have held powerful meanings for Middle Woodland peoples, and it should not surprise us that they would invoke them in the context of mediating the most profound social transformations the Eastern Woodlands had ever seen.

Acknowledgments

We would like to thank Nicole Boivin and Mary Ann Owoc for very critical, but very helpful editorial comments. Nicole in particular was very patient. Any remaining shortcomings are ours. We would also like to thank Ken Farnsworth and the Center for American Archeology for providing photographs. Geoarchaeological study of the Mound House was supported by the National Science Foundation under Grant No SBR-9729956.

References

Anderson, DG and Mainfort, RC, Jr (eds) (2002) *The Woodland Southeast*, Tuscaloosa: University of Alabama Press

Asch, DL and Asch, NB (1985) 'Prehistoric plant cultivation in west-central Illinois', in Ford, RI (ed), *Prehistoric Food Production in North America*, Museum Anthropology Papers No 75, Ann Arbor: University of Michigan, pp 149–203

Atwater, C (1820) *Description of the Antiquities Discovered in the State of Ohio and Other Western States*, Transactions and Collections of the American Antiquarian Society 1, Worcester: American Antiquarian Society

Baby, RS (1971) 'Prehistoric architecture: a study of house types in the Ohio valley' 71(4) *Ohio Journal of Science* 193–98

Braun, DP (1979) 'Illinois Hopewell burial practices and social organization: a reexamination of the Klunk-Gibson mound group', in Brose, DS and Greber, NB (eds), *Hopewell Archaeology: The Chillicothe Conference*, Kent: Kent State University Press, pp 68–79

Braun, DP (1987) 'Coevolution of sedentism, pottery technology, and horticulture in the central Midwest, 200 BC–AD 600', in Keegan, WF (ed), *Emergent Horticultural Economies of the Eastern Woodlands*, Center for Archaeological Investigations Occasional Papers No 7, Carbondale: Southern University of Illinois at Carbondale, pp 153–81

Braun, DP and Plog, S (1982) 'Evolution of "tribal" networks: theory and prehistoric North American evidence' 47 *American Antiquity* 504–25

Brose, DS (1979) 'A speculative model of the role of exchange in the prehistory of the Eastern Woodlands', in Brose, DS and Greber, NB (eds), *Hopewell Archaeology: The Chillicothe Conference*, Kent: Kent State University Press, pp 3–8

Brose, DS (1985) 'The Woodland Period', in Brose, DS, Brown, JA and Penney, DW (eds), *Ancient Art of the American Woodland Indians*, New York: Harry N Abrams, pp 43–92

Brose, DS and Greber, NB (eds) (1979) *Hopewell Archaeology: The Chillicothe Conference*, Kent: Kent State University Press

Brown, JA (1971) 'The dimensions of status in the burials at Spiro', in Brown, JA (ed), *Approaches to the Social Dimensions of Mortuary Practices*, Memoirs No 25, Salt Lake City: Society for American Archaeology, pp 92–112 (also issued as 36 *American Antiquity* [3, Pt 2], 92–112)

Brown, JA (1979) 'Charnel houses and mortuary crypts: disposal of the dead in the Middle Woodland period', in Brose, DS and Greber, NB (eds), *Hopewell Archaeology: The Chillicothe Conference*, Kent: Kent State University Press, pp 211–19

Brown, JA (1981) 'The search for rank in prehistoric burials', in Chapman, R, Kinnes, I and Randsborg, K (eds), *The Archaeology of Death*, Cambridge: CUP, pp 25–37

Brown, JA (1985) 'Long-term trends to sedentism and emergence of complexity in the American midwest', in Price, TD and Brown, JA (eds), *Prehistoric Hunter-Gatherers: The Emergence of Cultural Complexity*, Orlando: Academic, pp 201–31

Brown, JA and Vierra, RK (1983) 'What happened in the Middle Archaic? Introduction to an ecological approach to Koster site archaeology', in Phillips, JL and Brown, JA (eds), *Archaic Hunters and Gatherers in the American Midwest*, New York: Academic, pp 165–95

Buikstra, JE (1976) *Hopewell in the Lower Illinois Valley: A Regional Study of Human Biological Variability and Prehistoric Mortuary Behavior*, Scientific Papers 2, Evanston: Northwestern University Archeological Program

Buikstra, JE (1984) 'The lower Illinois River region: a prehistoric context for the study of ancient diet and health', in Cohen, NN and Armelagos, GJ (eds), *Paleopathology at the Origins of Agriculture*, Orlando: Academic, pp 215–34

Buikstra, JE and Charles, DK (1999) 'Centering the ancestors: cemeteries, mounds, and sacred landscapes of the ancient North American midcontinent', in Ashmore, W and Knapp, AB (eds), *Archaeologies of Landscape: Contemporary Perspectives*, Malden: Blackwell, pp 201–23

Buikstra, JE, Charles, DK and Rakita, GFM (1998) *Staging Ritual: Hopewell Ceremonialism at the Mound House Site, Greene County, Illinois*, Kampsville Studies in Archeology and History No 1, Kampsville: Center for American Archeology

Buikstra, JE, Charles, DK, O'Briant, KMH and Smith, KT (1997) *The Mound House Site: Summary of the 1995–1996 Investigations Conducted by the University of Chicago, the University of New Mexico and Wesleyan University*, Center for American Archeology, submitted to Illinois Historic Preservation Agency, Springfield

Bullington, J (1988) 'Middle Woodland mound structure: social implications and regional context', in Charles, DK, Leigh, SR and Buikstra, JE (eds), *The Archaic and Woodland Cemeteries at the Elizabeth Site in the Lower Illinois Valley*, Kampsville Archeological Center Research Series Vol 7, Kampsville: Center for American Archeology, pp 218–41

Butzer, KW (1977) *Geomorphology of the Lower Illinois Valley as a Spatial-Temporal Context for the Koster Site Archaic*, Reports of Investigations 34, Springfield: Illinois State Museum

Byers, AM (1998) 'Is the Newark Circle-Octagon the Ohio Hopewell "Rosetta Stone"?', in Mainfort, RC, Jr and Sullivan, LP (eds), *Ancient Earthen Enclosures of the Eastern Woodlands*, Gainesville: University Press of Florida, pp 135–53

Caldwell, JR (1964) 'Interaction spheres in prehistory', in Caldwell, JR and Hall, RL (eds), *Hopewellian Studies*, Scientific Papers Vol 12, Springfield: Illinois State Museum, pp 133–43

Charles, DK (1992) 'Woodland demographic and social dynamics in the American Midwest' 24 *World Archaeology* 175–97

Charles, DK (1995) 'Diachronic regional social dynamics: mortuary sites in the Illinois valley/American Bottom region', in Beck, LA (ed), *Regional Approaches to Mortuary Analysis*, New York: Plenum, pp 77–99

Charles, DK and Buikstra, JE (1983) 'Archaic mortuary sites in central Mississippi drainage: distribution, structure, and behavioral implications', in Phillips, JL and Brown, JA (eds), *Archaic Hunters and Gatherers in the American Midwest*, New York: Academic, pp 117–45

Charles, DK and Buikstra, JE (2002) 'Siting, sighting, and citing the dead', in Silverman, H and Small, DB (eds), *The Space and Place of Death*, Archaeological Papers of the American Anthropological Association No 11, Arlington: American Anthropological Association, pp 13–26

Charles, DK and Buikstra, JE (eds) (forthcoming) *Recreating Hopewell*, Gainesville: University Press of Florida

Charles, DK, Buikstra, JE and Konigsberg, LW (1986) 'Behavioral implications of the Terminal Archaic and Early Woodland mortuary practices in the lower Illinois valley', in Farnsworth, KB and Emerson, TE (eds), *Early Woodland Archeology*, Kampsville Seminars in Archeology No 2, Kampsville: Center for American Archeology, pp 458–74

Charles, DK, Leigh, SR and Buikstra, JE (eds) (1988) *The Archaic and Woodland Cemeteries at the Elizabeth Site in the Lower Illinois Valley*, Kampsville Archeological Center Research Series Vol 7, Kampsville: Center for American Archeology

Cole, F-C and Deuel, T (1937) *Rediscovering Illinois*, Chicago: University of Chicago Press

Conner, MD (1990) 'Population structure and skeletal variation in the Late Woodland of west-central Illinois' 82(1) *American Journal of Physical Anthropology* 31–43

Connolly, RP (1998) 'Architectural grammar rules at the Fort Ancient hilltop enclosure', in Mainfort, RC, Jr, and Sullivan, LP (eds), *Ancient Earthen Enclosures of the Eastern Woodlands*, Gainesville: University Press of Florida, pp 85–113

Count, EW (1952) 'The Earth-Diver and the Rival Twins: a clue to the time correlation in North-Eurasiatic and North American mythology', in Tax, S (ed), *Indian Tribes of Aboriginal America: Selected Papers of the XXIXth International Congress of Americanists*, Chicago: University of Chicago Press, pp 55–62

Dancey, WS and Pacheco, PJ (eds) (1997) *Ohio Hopewell Community Organization*, Kent: Kent State University Press

DeBoer, WE (1997) 'Ceremonial centres from the Cayapas (Esmeraldas, Ecuador) to Chillicothe (Ohio)' 7(2) *Cambridge Archaeological Journal* 225–53

Deuel, T (1952) 'The Hopewellian community', in Deuel, T (ed), *Hopewellian Communities in Illinois*, Scientific Papers 5, Springfield: Illinois State Museum, pp 249–65

Farnsworth, KB and Asch DL (1986) 'Early Woodland chronology, artifact styles, and settlement distribution in the lower Illinois valley region', in Farnsworth, KB and Emerson, TE (eds), *Early Woodland Archeology*, Kampsville Seminars in Archeology No 2, Kampsville: Center for American Archeology, pp 326–457

Farnsworth, KB and Wiant, MD (eds) (forthcoming) *Gregory Perino's Excavations at Certain Hopewellian and Late Woodland Sites in West-central Illinois*, Studies in Archaeology 4, Urbana: University of Illinois Press

Fie, SM (2000) *An Integrative Study of Ceramic Exchange During the Illinois Valley Middle Woodland Period*, unpublished PhD thesis, Department of Anthropology, State University of New York at Buffalo

Fowke, G (1905) 'The Montezuma Mounds' 2(5) *Missouri Historical Society* 1–16

Fowler, ML (1978) 'Cahokia and the American Bottom: settlement archeology', in Smith, BD (ed), *Mississippian Settlement Patterns*, New York: Academic, pp 455–78

Greber, NB (1979) 'A comparative study of site morphology and burial patterns at Edwin Harness Mound and Seip Mounds 1 and 2', in Brose, DS and Greber, NB (eds), *Hopewell Archaeology: The Chillicothe Conference*, Kent: Kent State University Press, pp 27–38

Greber, NB, Griffin, JB, Smart, TL, Ford, RI, Shane, OC III, Baby, RS, Langlois, SM, Belovich, SJ, Morse, DR and Vickery, KD (1983) 'Recent excavations at the Edwin Harness Mound, Liberty Works, Ross County, Ohio', *Midcontinental Journal of Archaeology*, Special Paper 5

Griffin, JB (1943) *The Fort Ancient Aspect: Its Cultural and Chronological Position in Mississippi Valley Archaeology*, Museum of Anthropology, Anthropological Papers 28, Ann Arbor: University of Michigan

Hajic, ER (1985) 'Landscape evolution and archeological contexts in the lower Illinois River valley' 5(2) *American Archeology* 127–36

Hajic, ER (1990) *Late Pleistocene and Holocene Landscape Evolution, Depositional Subsystems, and Stratigraphy in the Lower Illinois River Valley*, unpublished PhD thesis, Department of Geology, University of Illinois at Urbana-Champaign

Hall, RL (1977) 'An anthropomorphic perspective for eastern United States prehistory' 42(4) *American Antiquity* 499–518

Hall, RL (1979) 'In search of the ideology of the Adena-Hopewell climax', in Brose, DS and Greber, NB (eds), *Hopewell Archaeology: The Chillicothe Conference*, Kent: Kent State University Press, pp 258–65

Hall, RL (1997) *An Archaeology of the Soul: North American Indian Belief and Ritual*, Urbana: University of Illinois Press

Harrod, HL (1987) *Renewing the World: Plains Indian Religion and Morality*, Tucson: University of Arizona Press

Hassen, H and Farnsworth, KB (1987) *The Bullseye Site: A Floodplain Archaic Mortuary Site in the Lower Illinois River Valley*, Report of Investigations, No 2, Springfield: Illinois State Museum

Helms, MW (1988) *Ulysses' Sail: An Ethnographic Odyssey of Power, Knowledge, and Geographical Distance*, Princeton: Princeton University Press

Helms, MW (1995) *Creation of the Rainbow Serpent: Polychrome Ceramic Designs from Ancient Panama*, Albuquerque: University of New Mexico Press

Henderson, JG (1884) 'Aboriginal remains near Naples, Ill', in *Annual Report (1883) of the Smithsonian Institution*, Washington: Government Printing Office, pp 686–721

Hertz, R (1960) 'A contribution to the study of the collective representation of death', in *Death and the Right Hand*, Needham, R and Needham, C (trans), New York: Free Press

Hesselberth, C (1945) 'An Ohio type Hopewell ceremonial mound' 3(1) *Illinois State Archaeological Society Journal* 18–20

Hewitt, J (1983) *Optimal Foraging Models for the Lower Illinois Valley*, unpublished PhD thesis, Department of Anthropology, Northwestern University

Hudson, C (1976) *The Southeastern Indians*, Knoxville: University of Tennessee Press

Hughes, RE, Berres, TE and Farnsworth, KB (1998) 'Revision of Hopewellian trading patterns in midwestern North America based on mineralogical sourcing' 13 *Geoarchaeology* 701–30

Kut, ST and Buikstra, JE (1998) 'Calibration of C-14 dates in the lower Illinois River valley', paper presented at the 63rd Annual Meeting of the Society for American Archaeology, Seattle, Washington

Leeming, DA and Page, J (1998) *The Mythology of Native North America*, Norman: University of Oklahoma Press

Leigh, SR, Charles, DK and Albertson, DG (1988) 'Middle Woodland component', in Charles, DK, Leigh, SR and Buikstra, JE (eds), *The Archaic and Woodland Cemeteries at the Elizabeth Site in the Lower Illinois Valley*, Kampsville Archeological Center Research Series Vol 7, Kampsville: Center for American Archeology, pp 41–84

Lepper, BT (1996) 'The Newark Earthworks and the geometric enclosures of the Scioto valley: connections and conjectures', in Pacheco, PJ (ed), *A View from the Core: A Synthesis of Ohio Hopewell Archaeology*, Columbus: Ohio Archaeological Council, pp 224–41

Mainfort, RC, Jr (1988) 'Middle Woodland ceremonialism at Pinson Mounds, Tennessee' 53(1) *American Antiquity* 158–73

Mainfort, RC, Jr (1996) 'Pinson Mounds and the Middle Woodland period in the Midsouth and lower Mississippi valley', in Pacheco, PJ (ed), *A View from the Core: A Synthesis of Ohio Hopewell Archaeology*, Columbus: Ohio Archaeological Council, pp 370–91

McClintock, W (1999) *The Old North Trail, or Life, Legends, and Religion of the Blackfeet Indians*, Lincoln: University of Nebraska Press (reprint of 1910 edn)

Meltzer, DJ (1985) 'North American archaeology and archaeologists, 1879–1934' 50(2) *American Antiquity* 249–60

Metcalf, P and Huntington, R (1991) *Celebrations of Death: The Anthropology of Mortuary Ritual*, 2nd edn, Cambridge: CUP

Mills, WC (1904) 'Exploration of the Gartner mound and village site' 13 *Ohio Archaeological and Historical Quarterly* 128–89

Moorehead, WK (1899) 'Report of fieldwork in various portions of Ohio' 7 *Ohio Archaeological and Historical Quarterly* 110–203

Morrow, CA (1991) 'Observations on the Baehr Mound chert disks' 3 *Illinois Archaeology* 77–92

Neusius, SW (1986) 'Generalized and specialized resource utilization during the Archaic period: implications of the Koster site faunal record', in Neusius, SW (ed), *Foraging, Collecting, and Harvesting: Archaic Period Subsistence and Settlement in the Eastern Woodlands*, Center for Archaeological Investigations Occasional Papers No 6, Carbondale: Southern Illinois University at Carbondale, pp 117–43

Pacheco, PJ (ed) (1996) *A View From the Core: A Synthesis of Ohio Hopewell Archaeology*, Columbus: Ohio Archaeological Council

Perino, GH (1968) 'The Pete Klunk mound group, Calhoun County, Illinois: the Archaic and Woodland occupations (with an appendix on the Gibson mound group)', in *Hopewell and Woodland Site Archaeology in Illinois*, Bulletin 6, Urbana: Illinois Archaeological Survey, pp 9–124

Perino, GH (1973) 'The Late Woodland component at the Pete Klunk site, Calhoun County, Illinois', in *Late Woodland Archaeology in Illinois: Investigations in South-central Illinois*, Bulletin 9, Urbana: Illinois Archaeological Survey, pp 58–89

Riordan, RV (1998) 'Boundaries, resistance, and control: enclosing the hilltops of Middle Woodland Ohio', in Mainfort, RC, Jr and Sullivan, LP (eds), *Ancient Earthen Enclosures of the Eastern Woodlands*, Gainesville: University Press of Florida, pp 68–84

Rubey, WW (1952) *Geology and Mineral Resources of the Hardin and Brussels Quadrangles (in Illinois)*, United States Geological Survey, Professional Paper 218, Washington: Government Printing Office

Ruby, BJ (1997) *The Mann Phase: Hopewellian Subsistence and Settlement Adaptations in the Wabash Lowlands of Southwestern Indiana*, unpublished PhD thesis, Department of Anthropology, Indiana University

Schiffer, MB (1972) 'Archaeological context and systemic context' 37(2) *American Antiquity* 156–65

Schiffer, MB (1987) *Formation Processes of the Archaeological Record*, Albuquerque: University of New Mexico Press

Schlesier, KH (1987) *The Wolves of Heaven: Cheyenne Shamanism, Ceremonies, and Prehistoric Origins*, Norman: University of Oklahoma Press

Seeman, MF (1979) *The Hopewell Interaction Sphere: The Evidence for Interregional Trade and Structural Complexity*, Prehistoric Research Series 5(2), Indianapolis: Indiana Historical Society

Seeman, MF (1995) 'When words are not enough: Hopewell interregionalism and the use of material symbols at the GE Mound', in Nassaney, MS and Sassaman, KE (eds), *Native American Interactions: Multiscalar Analyses and Interpretations in the Eastern Woodlands*, Knoxville: University of Tennessee Press, pp 123–43

Silverberg, R (1968) *The Moundbuilders*, Greenwich: New York Graphic Society

Smith, KT, Van Nest, J, Charles, DK and Buikstra, JE (1997) 'Walls of sod: a new interpretation of Hopewell mound construction', paper presented at the 62nd Annual Meeting of the Society for American Archaeology, Nashville, Tennessee

Squier, EG and Davis, EH (1848) *Ancient Monuments of the Mississippi Valley*, Smithsonian Contributions to Knowledge 1, Washington: Smithsonian Institution Press

Staab, ML (1984) *Peisker: An Examination of Middle Woodland Site Function in the Lower Illinois Valley*, unpublished PhD thesis, Department of Anthropology, University of Iowa

Stafford, BD and Sant, MB (eds) (1985) *Smiling Dan: Structure and Function at a Middle Woodland Settlement in the Illinois Valley*, Kampsville Archeological Center Research Series Vol 2, Kampsville: Center for American Archeology

Struever, S (1960) *The Kamp Mound Group and a Hopewell Mortuary Complex in the Lower Illinois Valley*, unpublished MA thesis, Department of Anthropology, Northwestern University

Struever, S (1964) 'The Hopewell Interaction Sphere in riverine-western Great Lakes culture history', in Caldwell, JR and Hall, RL (eds), *Hopewellian Studies*, Scientific Papers Vol 12, Springfield: Illinois State Museum, pp 85–106

Struever, S (1968) 'Woodland subsistence settlement systems in the lower Illinois valley', in Binford, SR and Binford, LR (eds), *New Perspectives in Archaeology*, Chicago: Aldine, pp 285–312

Struever, S and Houart, GL (1972) 'An analysis of the Hopewell Interaction Sphere', in Wilmsen, EN (ed), *Social Exchange and Interaction*, Museum of Anthropology, Anthropological Papers 46, Ann Arbor: University of Michigan, pp 47–79

Styles, BW (1981) *Faunal Exploitation and Resource Selection: Early Late Woodland Subsistence in the Lower Illinois Valley*, Scientific Papers 3, Evanston: Northwestern University Archeological Program

Styles, TR (1985) *Holocene and Late Pleistocene Geology of the Napoleon Hollow Site in the Lower Illinois Valley*, Kampsville Archeological Center Research Series Vol 5, Kampsville: Center for American Archeology

Tainter, JA (1977) 'Woodland social change in west-central Illinois' 2(1) *Midcontinental Journal of Archaeology* 67–98

Thomas, C (1894) *Report on the Mound Explorations of the Bureau of Ethnology*, Twelfth Annual Report of the Bureau of Ethnology to the Secretary of the Smithsonian Institution 1890–91, Washington: Smithsonian Institution

Titterington, PF (1935) 'Certain bluff mounds of western Jersey County, Illinois' 1(1) *American Antiquity* 6–46

Titterington, PF (1950) 'Some non-pottery sites in St Louis area' 1 *Journal of the Illinois State Archaeological Society* 19–31

Van Gilder, C and Charles, DK (2003) 'Archaeology as cultural encounter: the legacy of Hopewell', in Jeske, RJ and Charles, DK (eds), *Theory, Method, and Practice in Modern Archaeology*, Westport: Praeger, pp 114–29

Van Nest, J (1997) *Late Quaternary Geology, Archeology, and Vegetation in West-Central Illinois: A Study in Geoarcheology*, unpublished PhD thesis, Department of Geology, University of Iowa

Van Nest, J (forthcoming) 'Rediscovering this earth: some ethnogeological aspects of the Illinois valley Hopewell mounds', in Charles, DK and Buikstra, JE (eds), *Recreating Hopewell*, Gainesville: University Press of Florida

Van Nest, J, Charles, DK, Buikstra, JE and Asch, DL (2001) 'Sod blocks in Illinois Hopewell mounds' 66(4) *American Antiquity* 633–50

Wiant, MD and McGimsey, CR (eds) (1986) *The Woodland Occupations of the Napoleon Hollow Site in the Lower Illinois Valley*, Kampsville Archeological Center Research Series Vol 6, Kampsville: Center for American Archeology

Wray, DE and MacNeish, RS (1961) *The Hopewellian and Weaver Occupations of the Weaver Site, Fulton County, Illinois*, Scientific Papers Vol 7 No 2, Springfield: Illinois State Museum

– 4 –

EARTH, WOOD AND FIRE

MATERIALITY AND STONEHENGE

Mike Parker Pearson

INTRODUCTION

In 1994 Colin Richards suggested he and I write a book, *Elemental Archaeology*, in which we explore the symbolic and practical dimensions of the elements – water, earth, stone, air, fire – through archaeological and ethnoarchaeological examples. We even found a publisher, but the years passed and nothing happened. The stresses and strains of increasing university workloads during the 1990s and the demands of our existing research commitments left little time for this embryonic idea to take off. Colin did, however, publish a paper on the significance of water for the siting of Neolithic henge monuments (1996) and the visit of my archaeological colleague Ramilisonina from Madagascar sparked for me an interest in the materiality of stone at Stonehenge and Avebury (Parker Pearson and Ramilisonina 1998). Other archaeologists have since experimented with ideas about materiality (Tilley 1999), examined the role of colour and texture of stone in Neolithic monuments (Tilley 2004), investigated soils as consciously modified material culture (Boivin 2000; Owoc 2001), and explored the role of fire as an encultured phenomenon (Moore 1997).

Ten years later, much has developed in the phenomenological study of archaeological elements and there have been some excellent and thought-provoking case studies, particularly of 'minerality' and 'vegetality' (see, for example, Rival 1998). Yet the surface has only been scratched, so to speak. This is perhaps all the more surprising given that many archaeologists are deeply aware of the richness of meaning which surrounds their source material – 'dirt'. We shouldn't need to go to Rajasthan to find out about the complex symbolism of mud plastering (Boivin 2000), or to Papua New Guinea to find out that people cover their bodies in ash at funerals (Strathern 1971). Nor should we need to go to Madagascar to discover that tomb dirt has properties of fertility or that travellers may take a small packet of soil from home when travelling afar (Mack 1986). There is plenty of symbolism in Western conceptions of dirt, that wide category of matter that incorporates various types of apparently mundane material. Within the household, it enters clean spaces on the soles of shoes; it exits bodily orifices, is trapped under fingernails or behind the ears, and comes out of the fireplace onto the carpet. Dirt is the focus of secular rituals in which even the most laid-back families may have rules about its containment. On the one hand, certain types of dirt can have religious connotations – for example, on Ash Wednesday when ash is marked on Christians' foreheads as a sign of repentance and purity. On the other, it is an ostensibly 'pragmatic' concern, as illustrated by the way that the zealous over-cleaning of dirt from house interiors serves as a measure of house-proud self-respect, supporting a multi-million pound industry in products and equipment.

Yet, as Fernand Braudel (1980) pointed out, we visit a foreign country to find out about that place but inevitably find out more about our own. It is many years since Mary Douglas made her famous pronouncement that dirt is simply 'matter out of place' (1966) but too few archaeologists have attempted to think seriously about the cultural significance of dirt, ash, earth, soil and sediment for the past societies that we study. Most recently, John Evans has written an environmental archaeologist's manifesto in which he demands recognition that 'soil and land texture are important referents for social expression ... not just for their functional attributes of fertility and knowledge but also as a means through which people communicated with each other' (Evans 2003: 71). His exhortations to environmental archaeological colleagues to appreciate past understandings of 'textures' in land and soil are supported with empirical evidence, and there are other examples that may be cited from British prehistory.

There are indications that minerality had profound significance for prehistoric peoples, just as it has for historical and contemporary societies. For example, there are a growing number of cases in which soil can be shown to have been brought from great distances to be incorporated into prehistoric settlements, defying rational explanation. Whereas the transport of Purbeck stone or Carrara marble over surprisingly long distances in the historical period makes sense in terms of the power of aesthetic values, the movement of soil is at first sight rather more perplexing.

During excavations at the Iron Age broch of Dun Vulan on South Uist in the Western Isles of Scotland, we noticed that the gallery within its 4 m-wide wall had been packed with an acidic soil different from the calcareous sand of the broch's locality on South Uist's west coast (Parker Pearson and Sharples 1999: 92). Thanks to the clearly defined vegetational geography of South Uist, it was possible to identify the specific zone from which this packing soil had derived on the basis of its palynological signature. The pollen in the soil derived not from the locality's coastal vegetation but from a narrow strip of land about 2 km to the east. Ash and sherds had been mixed with the soil which may have come from a contemporary settlement somewhere to the east. The soil represented many cartloads and was, in practical terms, no better than soils in the vicinity for providing stability or weather-proofing within the broch. Some years later, during excavations at the nearby Late Bronze Age settlement of Cladh Hallan, we noticed that the foundation of its houses had commenced with a deposit of occupation debris that had not formed *in situ*, but appeared to have been brought from another place of habitation (Parker Pearson *et al* 2004: 81).

In southern England, new work has thrown up further anomalies on Iron Age sites. East Chisenbury is a large tell-like mound of occupation debris that probably formed within two centuries between c 800 and 600 BC (Brown *et al* 1994). Within a single small trench, the excavators found a layer of Greensand soil (Macphail 1996) which occurs no closer than 15 km to the site. East Chisenbury is itself enigmatic, one of many Early Iron Age occupation mounds concentrated in the Vale of Pewsey, and currently defies conventional classification as a farmstead settlement. Further north in Oxfordshire, excavations of the Iron Age hillfort of Segsbury Camp have revealed that some of the chalk used in the rampart derives not from the immediate locality but from 2 km away, whilst limestone and sandstone inclusions were brought from an even greater distance (Lock and Gosden forthcoming). A similar incorporation of distant stone materials might also have been the case in the Middle Bronze Age ramparts of Ram's Hill (Bradley and Ellison 1975: 238; C Gosden, pers comm). New work on the gabbroic clays from the Lizard in Cornwall suggests that these had a symbolic significance beyond their functional suitability for making Bronze Age pottery (J Nowakowski, pers comm). In some cases, the raw clay appears to have been transported over much greater distances than is normally found in ethnographic cases (where the limit is usually 10–14 km; Arnold 1985: 32–57), with the finished vessel being composed of gabbroic clay mixed with local clay (J Wood, pers comm). Even before this observation was made, there was already a suspicion that the gabbroic clay – and perhaps more importantly the crushed gabbro inclusions – had a significance in prehistory far greater than can be explained by mere physical properties alone (Parker Pearson 1995).

THE MATERIALITY OF STONEHENGE

That visit by Ramilisonina to Stonehenge in February 1998 opened my eyes to the possibility that the stones at Stonehenge had been carefully selected for their material substance, the stone serving in contra-distinction to the contemporaneous 'woodhenges' 5 km upstream at Durrington Walls – timber post circles of similar geometry and size to Stonehenge (**Figures 4.1** and **4.2**). Others had already noticed the potential significance of different materials in this landscape (Bender 1998; Whittle 1997; see also Evans 2003). In particular, Barbara Bender independently drew analogies from Madagascar and Australia of stone representing persistence and immortality (1998: 44–45). She also suggested that growing things, 'things that are fertile', might have formed a counterpart to hard rock and white chalk. Following Julian Thomas' conception of chalk as an item of transaction between humans and the supernatural within a Neolithic economy of substances (Thomas 1997), she further emphasised that chalk was not only laid *on* the land but also dug *out of* it (Bender 1998: 47). Alasdair Whittle had also noted the potential significance of the ancestral permanence of stone at Stonehenge (1997).

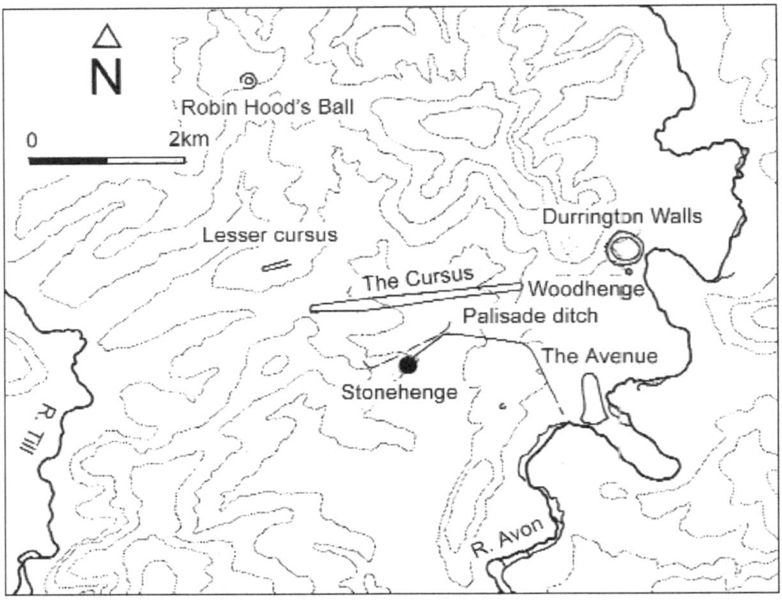

Figure 4.1: The major Neolithic monuments of the Stonehenge landscape (drawn by Irene de Luis).

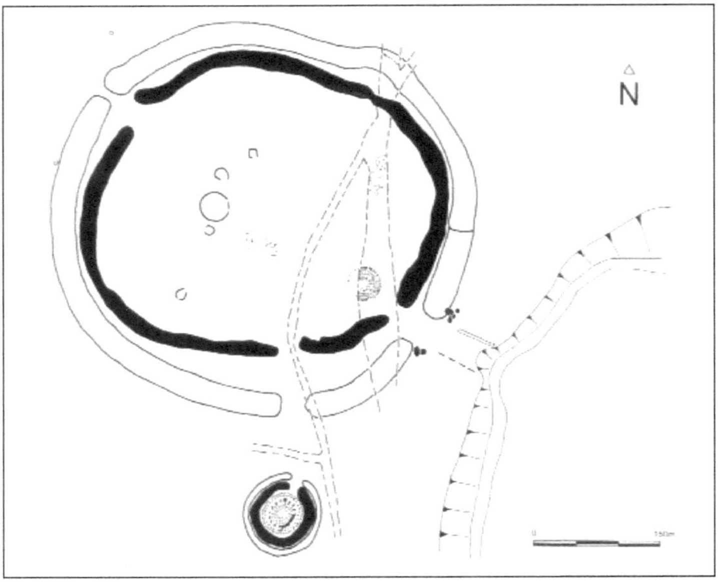

Figure 4.2: Durrington Walls and Woodhenge (drawn by Irene de Luis).

Wood and stone could be understood as set in simple binary opposition, but Ramilisonina's vision provided, more satisfyingly, an understanding of materials as elements within a metaphorical process of hardening – the passage from death into ancestorhood – which was represented not only in Stonehenge's 'lithification' (from a timber monument to a stone one) through time but also in the passage downstream from the wooden circles at Durrington Walls to the stone circle at Stonehenge. We speculated that Durrington Walls might have been the site of funerary feasting where bodies (in retrospect, more likely cremated ashes) were dispatched into the river, their spiritual elements being helped along the Stonehenge Avenue to reside as permanent, eternal ancestors within Stonehenge.

Francis Pryor extrapolated from the wood-stone theory to suggest that the Late Neolithic and Early Bronze Age barrows and cairns at Brenig in North Wales were also arranged according to a spatial progression from wood to stone (2003: 257–61), the monuments on the western side of the valley being constructed with earth and wooden posts in contrast to the stone cairns on the east, with those in-between the two groups forming a transitional group. Similarly, Mary Ann Owoc (2000) has noted a progression from wood enclosures to stone cairns or caps during the developmental history of particular round barrows in south-west Britain as the associated rites of passage progressed.

In 2003 Colin Richards and I assembled a group of Neolithic specialists to commence the Stonehenge Riverside project in order to follow up the wood-stone theory and improve our understanding of Stonehenge's landscape and social context (Parker Pearson *et al* 2004). Whilst that project has just begun, certain new insights have already been made as Colin and I have begun to explore both the landscape and the published literature on previous excavations within it, and this chapter provides an opportunity to present a few of them.

CHALK AND WOOD: DURRINGTON WALLS

The simple wood-stone theory fails to provide a developed view of the significance of earth and chalk, two other substances that were clearly of immense practical concern to Neolithic people, who came into regular contact with them through cultivation and monument construction. Around 2400 BC, there was a major phase of construction in which the sarsens were erected at Stonehenge (Cleal *et al* 1995) and, upstream, Durrington Walls – the largest henge in Britain – was built as a bank and ditch enclosing an area which contained circular arrangements of wooden posts (Wainwright with Longworth 1971). Whether the Durrington Walls henge was built at precisely the same time as the stones were erected at Stonehenge is not known. However, the juxtaposition of the two monuments at either end of a short stretch of the Avon River and the complementarity of the activities carried out within each suggest that they were built within the same programme to form a single inter-linked entity. Although ceremonial use of this landscape already went back at least a thousand years by this time, the sheer scale of this work programme suggests a millenarian zeal in which people from across Britain must have participated in a religiously inspired remodelling of cosmology.

This new complex highlighted certain astronomical orientations not hitherto marked in the monuments of Salisbury Plain. At Durrington Walls, visitors moved out of the southern timber circle through the east entrance on the axis of the midwinter solstice sunrise and out of the south entrance towards the henge annexe of Woodhenge, entering it on the midwinter solstice sunset axis. Downstream they would have been able to walk along the Avenue from the riverbank and enter Stonehenge in that same direction. Confirmation of the calendrical significance of this river-linked complex has come from analysis of the feasting debris left at Durrington Walls. Umberto Albarella and Dale Serjeantson have re-analysed the excavated animal bones and have found that the pigs, which had been barbecued for feasting, had been culled mostly in the winter period (Albarella and Serjeantson 2002; U Albarella, pers comm). I envisage the main events at Durrington Walls as midwinter feasts of the dead, accompanied by a series of smaller non-calendrical occasions such as funerals and other mortuary-related activities. In simpler words, activities at Stonehenge and Durrington Walls marked a kind of 'Neolithic Christmas'.

Whereas Stonehenge in 2400 BC was a masterpiece of stoneworking, Durrington Walls was an enormous earth sculpture in chalk. Its builders selected a steep dry valley and, despite the sloping ground, imposed a symmetrical, oval feature upon it, in the form of a massive chalk bank, parts of which still stand to about 8 m. The ditch for this bank was, like those of most henges, dug on the inside. Its scalloped edges reveal that it was gang-dug in 40 m sections to 6 m deep. Curiously, the ditch was not symmetrical in plan: for reasons unknown, there was a 30 m-wide berm[1] on the south side, astride the entrance towards Woodhenge (**Figure 4.2**). There might already have been activity within the dry valley before 2400 BC. Inside the east entrance this survives as a complex sequence of timber posthole arrays, mostly in circular forms (known as the Southern Circle; Wainwright with Longworth 1971). Near the north entrance there is another, the Northern Circle. Woodhenge was another of these post arrangements and there was a linear array of small posts outside the south entrance of Durrington Walls.

Many of the postholes excavated by Geoffrey Wainwright at Durrington Walls, such as those in the Southern Circle, were enormous, up to 2 m deep and over 1 m wide. The posts, presumably whole tree trunks, were set in the ground and then left until they rotted. At this point, perhaps 160 years later, people returned to dig pits into the top of these postholes and deposit offerings of pig bones, flint and bone tools and broken pottery before the pits filled up with a fine silt. In some cases, the sequence then started again, with someone digging out a new posthole next to an old one. In one or two cases, the silt-filled pit was re-cut. The origin of this silt is difficult to determine. It was initially thought to have been carried in gradually by natural processes of wind and rain as a colluvial deposit, but the pits' steep sides and the unweathered state of the sherds placed in them show that these features were swiftly filled.

The posthole sequences provide an interesting elaboration on the wood-stone theory since they support the notion of the transition as a metaphor of the life passage. The timbers had to be seen to decay, to the extent that their passing was marked by the digging of an offering pit. As metaphors of the post-mortem process of corruption, these tree trunks also embodied the memorialisation of the dead. The pits dug into the rotted post bases to receive closing offerings were then filled not with the extracted chalk rubble and rotted wood but with fine silt topsoil, a medium intermediate between living and dead things, between growing substances and chalk. In their transitory bodily form, the individual dead, represented by the rotted wood, would eventually be forgotten. Downstream their collective, ancestral memory would live forever as Stonehenge.

FIRE AND WATER: THE STONEHENGE RIVER

Between Durrington Walls and the beginning of the Stonehenge Avenue, the River Avon twists and turns, disorienting anyone who passes along it. Although we do not yet know the river's course four and a half thousand years ago, the positioning of chalk river cliffs suggests that it was similarly tortuous. Perhaps this served to confuse the spirits of the dead and prevent them from returning to haunt the living.[2] The valley bottom probably also presented Neolithic people with a dramatic floodplain landscape that contrasted with the solid chalk downland through which the River Avon flowed. Pollen and snail evidence indicates that this floodplain was a treeless, wet, sedge fen, a more or less permanent swamp whose unsteady surface contrasted with the surrounding chalk (Evans 2003: 80; Cleal *et al* 2004: 234). Flowing on past Durrington Walls, the Avon entered a gorge-like setting, lined on its west side by a high ridge, before turning westwards at the chalk river cliff of Ratfyn. High up on these cliffs on either side of this stretch of the river, people dug pits, not to hold posts, but to deposit some unusual materials, including wood ash and charcoal.

We have no idea how many ash pits were dug along the cliff tops of the river – the chance nature of the discoveries suggests that there are many more to be found. Four of them are known from the west bank south of Woodhenge at Woodlands (Stone and Young 1948; Stone 1949) and another was found on the east side of the river further downstream at Ratfyn, amongst a group of three other prehistoric features (Stone 1935a) (**Figure 4.3**). These steep-sided, flat-bottomed pits had been filled with wood ash and/or charcoal. Unburnt artefacts and animal bones had been mixed in, including flint tools and flakes, lumps of sarsen and flint, marine shells, wafer-thin, poorly fired and highly decorated pottery and (in the Ratfyn pit) a bear's scapula. Often more than a metre wide and half a metre deep, these pits contained single-episode dumps of ash and charcoal which must have been scooped up from extinguished fires. To accumulate wood ash in sufficient quantity to fill such pits requires the burning of many tons of wood, and these fills would seem to have been the product of large and long-lasting outdoor fires. One pit contained only a small deposit of

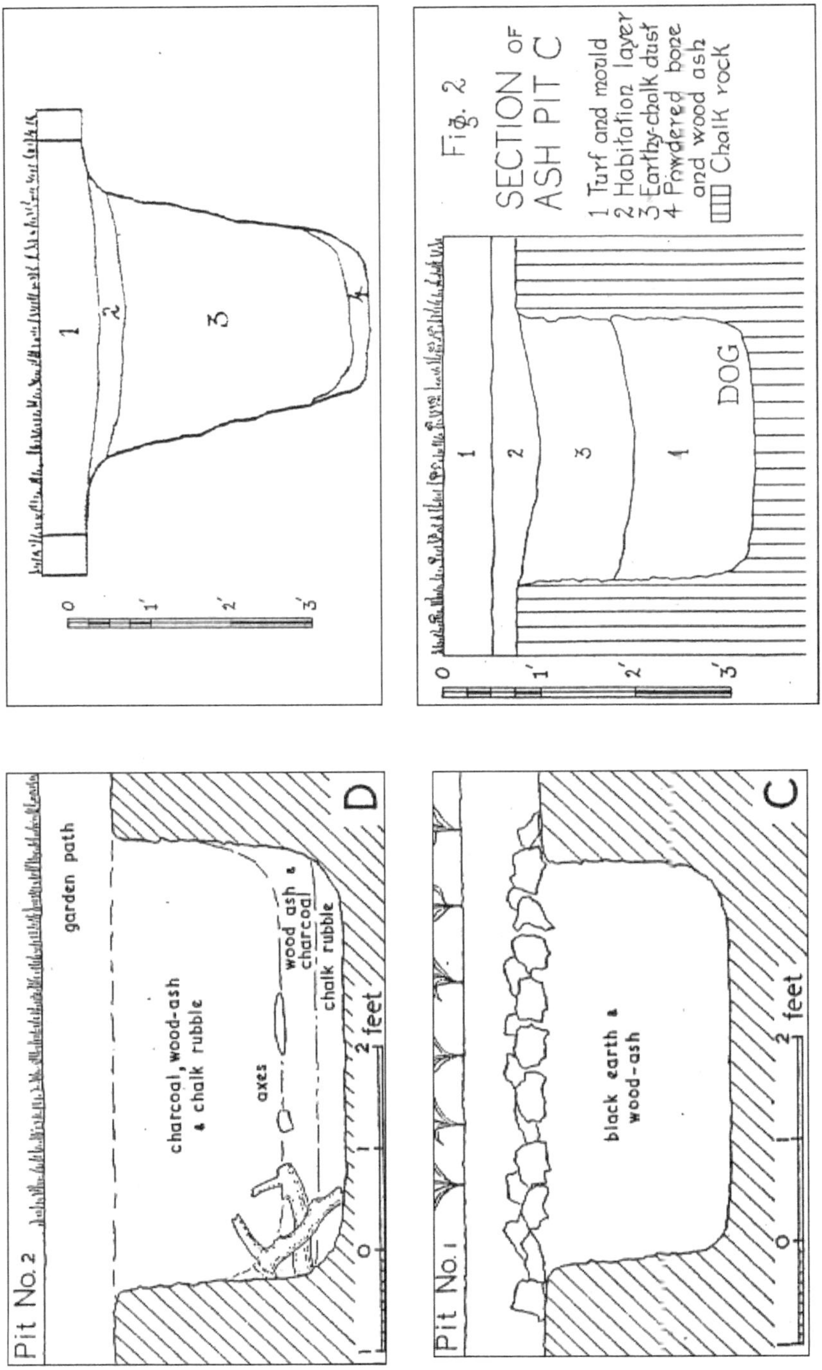

Figure 4.3: Ash pits from 'Woodlands' south of Woodhenge (left, top and bottom), Ratfyn (top right; layer 3 is the ash layer) and Easton Down (bottom right) drawn by JFS Stone (from Stone 1935a, 1935b and Stone and Young 1948; reproduced by kind permission of the Wiltshire Archaeological and Natural History Society).

wood ash and charcoal, dumped into a chalk dust and rubble fill. However, the shape of the deposit indicates that it had been carried in a flat-based basket.

It seems most likely that the ash pits are the residues of practices along the river cliffs involving the setting of huge fires, possibly at calendrical or funerary celebrations, and involving special ceramics important for their appearance and not their use as containers.[3] Furthermore, although this part of the Avon is 50 km from the sea, the deliberate inclusion of marine shell strikes a chord with the finds from Durrington Walls, where numbers of oysters were unearthed (Harcourt 1971: 338). Marine shells were also the most common temper added to the Grooved Ware found within Durrington Walls (Cleal 1995). Although shell has excellent thermal properties as a ceramic temper, its inclusion in Grooved Ware is unusually prominent at Durrington Walls. The placing of marine shells in these contexts perhaps recalled and invoked the bones and cremated ashes of the dead, as well as their eventual destination as they drifted downriver and ultimately out to sea. In this connection, the presence of sarsen lumps in two of the pits may also be significant. On one level, the burial of wood ash in circular, cylindrical pits might have been pragmatic since its high content of calcium carbonate increases the alkalinity of chalk soils to a level harmful for cultivation. On another, some of the items discarded with the ash in the pits – complete and broken flint axes, and elaborately decorated 'Woodlands' Grooved Ware (Stone 1935a, 1935b; Stone and Young 1948; Stone 1949) – are not casual discards.

The likely focus of ceremonial activities at midwinter makes the experience of such events a far cry from the invented summer solstice gatherings of recent years. At a dark, cold, wet and even snow-covered time of year, enormous and continuous bonfires, burning wood brought from many miles away, would have been foci for large gatherings. The kinds of ash deposits that are found indicate that it is more likely that the wood ash was produced by burning many tons of timber at such short-term events, rather than through the gradual collection of ash accumulated over many years. When we consider that the entire local landscape at this time was largely or wholly devoid of trees, the provisioning of huge quantities of firewood must have been a considerable logistical arrangement over long distances, at least on a par with the transport of timbers to build the 'woodhenges'.[4]

ASH AND TURF: THE ROUND BARROWS OF SALISBURY PLAIN

In Britain, in the period after 2400 BC until about 1700 BC, the dead (or at least some of them) began to be retained within the landscape, their remains often being marked by mounds known as round barrows. On Salisbury Plain, these barrow mounds are often formed largely or entirely from earth and chalk derived from the upcast of the circular ditches dug around them, heaped over a core of soil or turf (Ashbee 1960: 44). Yet there are some significant exceptions. On the Bulford army range halfway up the Nine Mile River, a tributary which feeds into the Avon close to Durrington Walls, William Hawley (who excavated at Stonehenge in 1919–26) dug a group of round barrows of which one (Bulford 3) was of unusual construction (1910: 618–20). Its outer layer of chalk was strongly compacted, but 2 m down, Hawley encountered the top of a 2 m-deep mass of wood ash – 'tons of it' (*ibid*: 619). Amongst this ash were the charred logs and vertical ash and oak timbers of a rectangular structure. Sherds and the heavily burnt fragments of a cremation were present and he surmised that there was burnt turf in the top of the ash layer. Yet this was no cremation deposit. Hawley was well aware how little ash is left by even the largest of wood fires and concluded that the evidence pointed to 'a fire having been kept up for a very long time' (*ibid*: 619). He abandoned the trench reluctantly when 'the ash collapsed and buried a man nearly to his shoulders' (*ibid*).

The significance of wood ash and charcoal as a deposit in a mound or in a pit might have been multiple. Not only might it mimic the transformation of a corpse into ashes (as was certainly practiced in the Late Neolithic), but it might also have represented a transition between the organic and the permanent, by which wood was reduced to its inorganic mineral form. Wood ash is also the by-product of those fires which perhaps lit the darkness at midwinter. Large, communal fire rituals might well have been orchestrated as acts to encourage the sun's return, to halt the annual descent into death and renew the world's life force.

The presence of ash and 'occupation debris' within round barrow mounds has been noted in many instances. Leslie Grinsell listed 15 in Wessex (1941: 105) and surmised that, whilst it is unusual for such foreign material to be brought to the locality, 'it may well have been brought there by mourners coming from a distance, each bringing basketloads for the mound' (*ibid*: 94). Paul Ashbee also noted the incorporation of dark occupation soil, containing sherds, charcoal and bone, in round barrows across Britain (1960: 55), whilst Mary Ann Owoc has explored the significance of the different soils incorporated into Bronze Age barrows in the south-west (2001).

Round barrows gradually filled up the landscape of Salisbury Plain on either side of the Avon, eventually forming the densest and largest grouping in Britain. The Nine Mile River group is a particularly dense component of the funerary landscape, as are the cemeteries in a doughnut-shaped distribution around Stonehenge (Woodward and Woodward 1996). The latter form impressive linear cemeteries, many of them located so as to be highly visible from Stonehenge. In four cases, each cemetery is positioned off one end of an earlier henge (G Lambrick, pers comm). One of these is King Barrow Ridge, with Coneybury henge to its south (**Figure 4.4**). The impressively large round barrows on this ridge occupy an important position within the Stonehenge environs. On walking up the line of the Avenue from the riverside, people passed through a series of linear barrow groups. Six hundred metres from the riverside, a first façade of barrows sits astride the Avenue. A second line of six round barrows lies 350 m further on. On gaining the ridge where Stonehenge finally becomes visible, 500 m further, people passed through the most impressive façade of 11 barrows on the left-hand side and 11 barrows on the right-hand side (**Figure 4.5**).

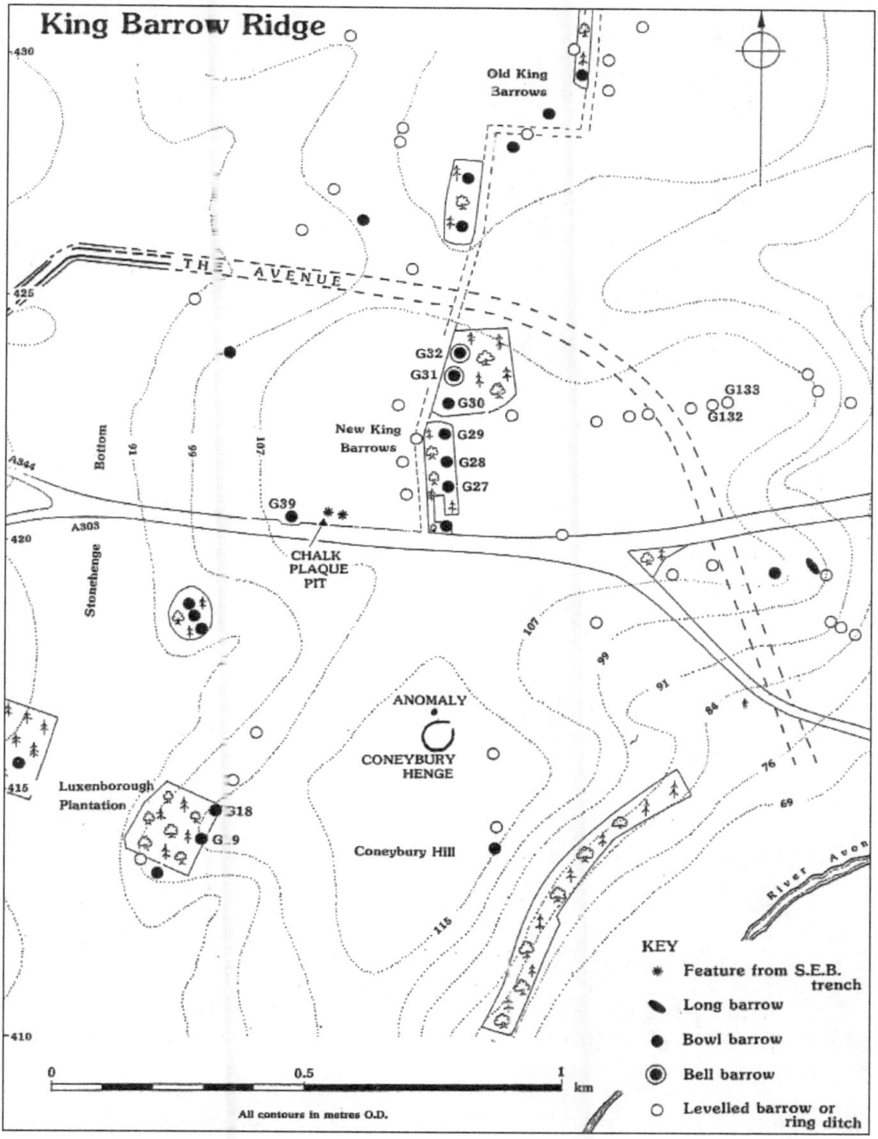

Figure 4.4: The round barrows forming façades along the Stonehenge Avenue (drawn by Liz James and reproduced by kind permission of Wessex Archaeology).

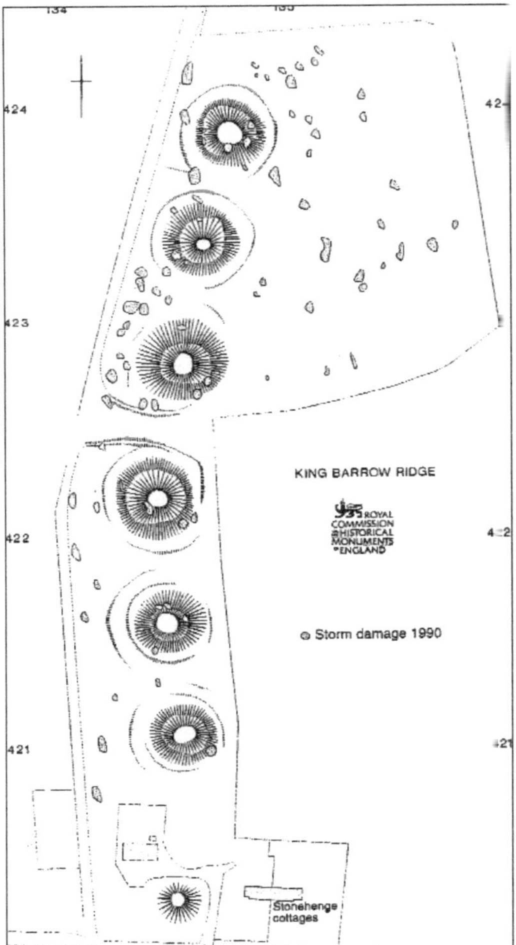

Figure 4.5: The round barrows on King Barrow Ridge, south of the Avenue (Crown copyright, reproduced by kind permission of the Secretary and Commissioners of RCHME).

The barrows on this ridge are the gatekeepers of Stonehenge and it is very possible that the graves they mark were those of the most important lineage or dynasty associated with the monument. No burials have ever been excavated within them because trees growing on top of them prevented antiquarians from digging into them in the 18th and 19th centuries. In 1987 and 1990, fierce storms brought down some of the trees and archaeologists had an opportunity to find out how six of the barrows on the south side of the Avenue were built (Cleal and Allen 1994: 57–60). Four were constructed out of a soil and turf stack capped

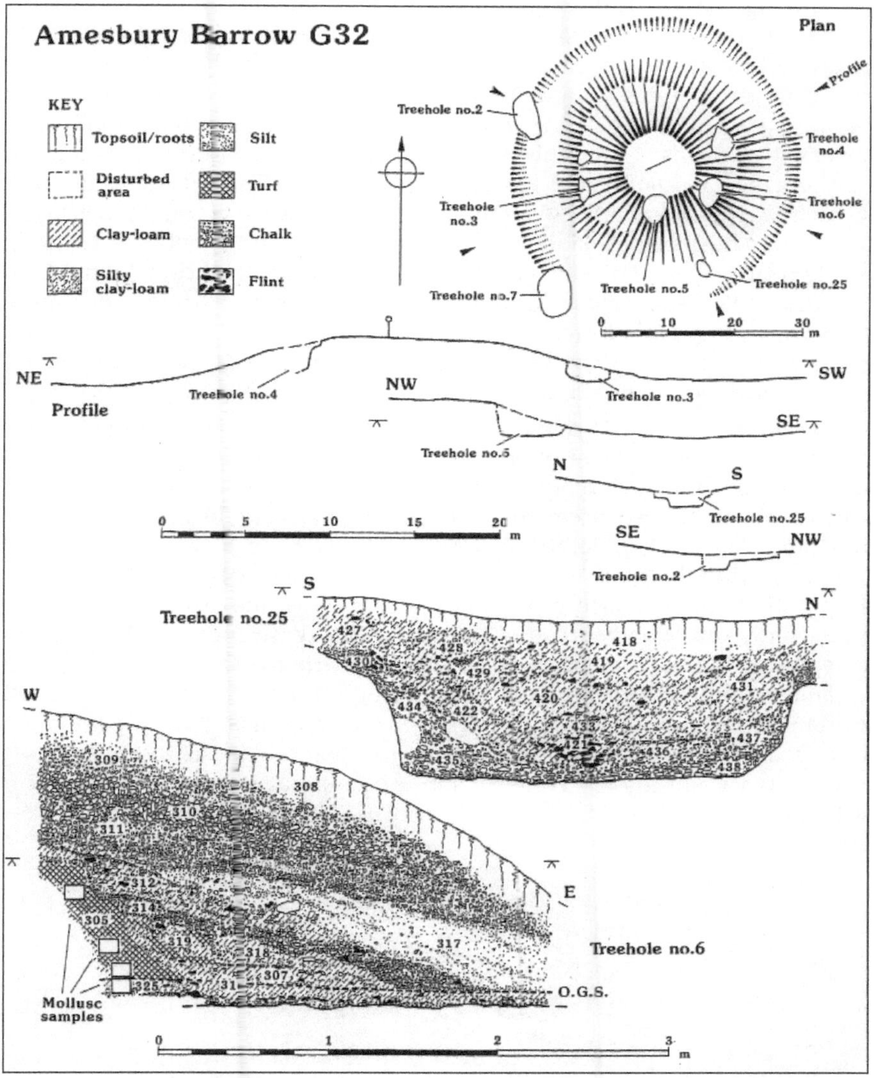

Figure 4.6: Sections from tree holes in Amesbury G32, one of the round barrows on King Barrow Ridge. Note the turf core, covered by loam, silt and a chalk capping (drawn by Liz James and reproduced by kind permission of Wessex Archaeology).

with chalk rubble from the surrounding ditch (**Figure 4.6**). The other two were constructed solely of soil and turf. In addition, another four ploughed-out barrows in the group also had turf and soil cores.

The King Barrow Ridge barrows are by no means the only ones in the Stonehenge area, let alone within Britain, to have been built predominantly or entirely with turf. However, this is a relatively rare method of construction which is unusual in this instance because of the number of barrows built in this fashion. It also had significant consequences for land use. Like much of the Stonehenge environs around this period, the vegetation on the barrow ridge was lightly grazed grassland, perhaps with some shrubs. The turves accumulated into each barrow originally covered an area of at least a hectare and probably much more. The decommissioning of grazing land covering well over 10 ha (and possibly nearer 30 ha if all the barrows on the ridge had turf cores) would have had a dramatic visual impact, simultaneously revealing a white chalk surface and depriving herders of grazing land. If the turves came from the immediate vicinity of the barrows, then they would have laid bare a great swathe of land through which the Avenue crossed.

How might we explain this wasteful stripping of turf? It might have been a form of conspicuous consumption, destroying pasture that formerly fed cattle, and echoing the destruction of the cattle herds themselves, a practice revealed by the cattle bones heaped upon round barrows in eastern England at this time (Davis and Payne 1993; Selkirk 2004). If cattle were to be destroyed in their hundreds during the funerary rituals, then why not also destroy the pastures at the same time, thereby also providing sustenance for the slaughtered herds in the afterlife? Ann Woodward has asked whether these presumably successive acts represent a drastic change in land use or, alternatively, were symbolic acts: 'the claiming and subsequent sealing or "hiding" of the whole land surface of an important sacred space' (2000: 52). John Evans has speculated that the removal of turf was an end in itself: 'it is as if people wanted to create visibility or monumentality not just by constructing mounds but by changing the whole distribution of fertility within the land' (2003: 84) (**Figure 4.7**).

Figure 4.7: Heaping up fertility into a round barrow. A mound of manure in southern Britain in the third millennium AD. Unlike barrows of 4,000 years ago, this is destined to be spread over the fields (© M Parker Pearson).

Additionally, these acts of landscape transformation might also have signified the transition from death into ancestorhood, mingling the dying grass with soil (that intermediate substance between living things and chalk) and revealing the bare chalk beneath, perhaps metaphorically the bones of the land. The topographical position of these barrows, midway between the riverside and Stonehenge, is also intriguing in this respect as a liminal space in which the semi-permanence of earth might have been contrasted with the permanence of chalk and stone and the transitory properties of formerly living things. This juxtaposition of materials soft and solid is further highlighted by the placing of an unusual deposit on King Barrow Ridge. In the central pit of one of the disc barrows there were not human bones but a collection of oddly shaped natural flints, 'each chosen apparently for its curious projecting knobs and legs', which 'must have taken years to bring together' (Passmore 1940). Perhaps these unusual flint nodules were perceived as bones of the most ancient ancestors, transformed into the flint that made daily work possible.

There are other barrows in the Stonehenge area where the stripping of soil to reveal the chalk was an important aspect of the ritual sequence. The corpse buried within Bush Barrow, to the south of Stonehenge within the Normanton Down barrow group, was laid directly on the surface of the chalk after the area to be covered by the mound had been cleared of turf and soil. The curious saucer and disc barrows on Salisbury Plain were also designed to permanently expose large expanses of bare chalk surrounded by a circular ditch.

Finally, chalk was a suitable medium for inscribing geometric designs and other motifs. There are interesting Neolithic and Early Bronze Age examples of this from Folkton, Yorkshire, and from elsewhere in Wessex such as Flagstones, Dorset. However, one of the best examples is a pair of incised chalk plaques found in a pit 200 m west of King Barrow Ridge (Vatcher 1969; Harding 1988; Cleal and Allen 1994: 60–62). In other parts of Britain, stone was occasionally used for such motifs but, other than Grooved Ware ceramics, no other media such as bone or antler seem to have been decorated. Of course, we have no idea whether textiles or skins were similarly decorated, but the absence of bone or antler carving potentially singles out chalk and other inorganic materials as special in this respect (cf Thomas 1996: 150).

CONCLUSION

> I garði þessum er haugr, hrœrt allt saman gull ok silfr ok mold.
>
> In the settlement there is a burial mound in which there is gold and silver and soil mixed together.
>
> (*Óláfs saga helga* Snorri Sturluson c 1178–1241, cited in DuBois 1999: 4)

Earth has often been considered by archaeologists to be a symbolically neutral, surrounding matrix in which finds are buried and with which features are filled. The examples of wood ash and charcoal soils from prehistoric pits and barrows, and also of soil and turf burial mounds, demonstrate that we must be prepared to recognise that their incorporation and construction were carefully considered acts. In particular, the deposition of wood ash – wood turned to mineral – gives us an interesting insight into the nature of midwinter fire rituals along the Avon, a probably treeless environment into which enormous quantities of firewood had to be brought for these events to ensure renewal of the sun's force. Even though we could consider the digging of pits to be a pragmatic means of disposing of this highly alkaline substance, wood ash was often buried with deposits that are most effectively interpreted as votive offerings. As the quotation from the Icelandic saga implies, the act of burial might have constituted not only a removal of things from the surface world, but also a mixing of those things with soil. We have to consider that pits were dug to be filled with matter that not only had to be removed from the above-ground world, but was also thereby permanently altered through its contact and mixing with other constituents.

The use of topsoil in filling offering pits and in heaping up over burials can be suggested to have been an important liminal act that served not only to remove land from potential agricultural activity but also to mark the transition from temporary vitality (wood and flesh) to ancestral permanence (stone, chalk, flint and bones). In this sense, the transition of timber through burning to wood ash also embodied this hardening. It is additionally possible that the calcareous trinity of ash, bone and chalk also represented the durable, whitened cores of the more ephemeral substances in which they were clothed or embedded. Thus they may have been conceived as the hard core that remained after the organic had been stripped away by time and rites of passage.

In conclusion, around 2400 BC, Stonehenge and Durrington Walls formed an integrated monument to a great midwinter festival lasting probably several months, in which the high point was the sun's rebirth, celebrated at the solstice, the liminal moment between living and dead when ancestors walked the earth. For northern European countries, this midwinter celebration remains the key calendrical festival despite its appropriation by Christianity,[5] thereby giving it a recognisable familiarity. Yet the ways in which timber, topsoil, turf, flint, stone and chalk were used and transformed are perhaps very alien to our modern world, reminding us that the past was indeed a different country, familiar only in barest outline.

Notes

1 A berm is the gap between a ditch and its upcast bank.

2 This disorientation of the dead is a common factor in many parts of the world, aimed at ensuring their successful separation from the living (Huntington and Metcalf 1979; Parker Pearson 1999: 25). It often takes the form of moving the dead across water or carrying the coffin on a winding journey along an indirect route.

3 Other small groups of pits, containing Beaker sherds, with at least one layer of charcoal dust or ash have been found on 14 published sites over a 400 sq km area around Stonehenge (Heaton and Cleal 2000: 79). At least one of these groups, three ash pits from Easton Down, contained substantial quantities of ash (Stone 1935b:

74). The 'Woodlands' sub-style of Grooved Ware is unusually thin – with its coarse fabric too thin to cook with – and its applied ribs are elaborately decorated rather than being plain as in most other Grooved Ware vessels.

4 It is not inconceivable that the above-ground sections of the rotted posts from the timber circles at Durrington Walls and Woodhenge were reused as firewood.

5 Compare the importance of Easter over Christmas in southern Europe; on the basis of the Christian message, Easter ought to be the most important festival throughout Christendom.

Acknowledgments

I thank Colin Richards for getting me started in 'elemental archaeology' and for encouraging us to set up the Stonehenge Riverside project. Dave Field, Chris Gosden and Dave McOmish provided important empirical information on soils whilst Jacqui Nowakowski kept me up to date with new discoveries about Bronze Age ceramics in Cornwall. Thanks also to Mike Allen for comments on barrow contents, George Lambrick for his observation on henges and barrow cemeteries, and Geoff Wainwright for his comments on pits dug into postholes at Durrington Walls. Umberto Albarella provided new unpublished results on pig seasonality at Durrington Walls. I am also grateful for information on Cornish clays from Jacqui Wood, delivered in a paper in her absence at the Prehistoric Ceramics conference in Bradford in 2002. Karen Godden edited and commented on the text. Further useful comments were received from the volume's editors.

References

Albarella, U and Serjeantson, D (2002) 'A passion for pork: meat consumption at the British Late Neolithic site of Durrington Walls', in Miracle, P and Milner, N (eds), *Consuming Passions and Patterns of Consumption*, Cambridge: McDonald Institute, pp 33–49

Arnold, D (1985) *Ceramic Theory and Cultural Process*, Cambridge: CUP

Ashbee, P (1960) *The Bronze Age Round Barrow in Britain*, London: Phoenix House

Bender, B (1998) *Stonehenge: Making Space*, London: Berg

Boivin, N (2000) 'Life rythyms and floor sequences: excavating time in rural Rajasthan and Neolithic Catalhoyuk' 31(3) *World Archaeology* 367–88

Bradley, R and Ellison, A (1975) *Ram's Hill*, Oxford: BAR British Series 19

Braudel, F (1980) *On History*, London: Weidenfeld and Nicolson

Brown, G, Field, D and McOmish, D (1994) 'East Chisenbury midden complex', in Fitzpatrick, AP and Morris, EL (eds), *The Iron Age in Wessex: Recent Work*, Salisbury: Trust for Wessex Archaeology

Cleal, RMJ (1995) 'Pottery fabrics in Wessex in the fourth to second millennia BC', in Kinnes, I and Varndell, G (eds), *'Unbaked Urns of Rudely Shape': Essays on British and Irish Pottery for Ian Longworth*, Oxford: Oxbow

Cleal, RMJ and Allen, M (1994) 'Investigation of tree-damaged barrows on King Barrow Ridge and Luxenborough Plantation, Amesbury' 87 *Wiltshire Archaeological and Natural History Magazine* 54–84

Cleal, RMJ, Allen, M and Newman, C (2004) 'An archaeological and environmental study of the Neolithic and later prehistoric landscape of the Avon valley and Durrington Walls environs' 97 *Wiltshire Archaeological and Natural History Magazine* 218–48

Cleal, RMJ, Walker, KE and Montague, R (1995) *Stonehenge in its Landscape: Twentieth-century Excavations*, London: English Heritage

Davis, S and Payne, S (1993) 'A barrow full of cattle skulls' 67 *Antiquity* 12–22

Douglas, M (1966) *Purity and Danger*, Harmondsworth: Penguin

DuBois, TA (1999) *Nordic Religions in the Viking Age*, Philadelphia: University of Pennsylvania Press

Evans, JG (2003) *Environmental Archaeology and the Social Order*, London: Routledge

French, CAI (2003) *Geoarchaeology in Action: Studies in Soil Micromorphology and Landscape Evolution*, London: Routledge

Grinsell, LV (1941) 'The Bronze Age round barrows of Wessex' 7 *Proceedings of the Prehistoric Society* 73–113

Harcourt, R (1971) 'The animal bones', in Wainwright, GJ (ed) with Longworth, IH, *Durrington Walls: Excavations 1966–1968*, London: Society of Antiquaries

Harding, P (1988) 'The chalk plaque pit, Amesbury' 54 *Proceedings of the Prehistoric Society* 320–27

Hawley, W (1910) 'Notes on barrows in south Wilts' 34 *Wiltshire Archaeological and Natural History Magazine* 615–28

Heaton, M and Cleal, RMJ (2000) 'Beaker pits at Crescent Copse, near Shrewton, Wiltshire, and the effects of arboreal fungi on archaeological remains' 93 *Wiltshire Archaeological and Natural History Magazine* 71–81

Huntington, R and Metcalf, P (1979) *Celebrations of Death: The Anthropology of Mortuary Ritual*, Cambridge: CUP

Lock, G and Gosden, C (forthcoming) *Segsbury Camp: Excavations in 1996 and 1997 at an Iron Age Hillfort on the Oxfordshire Ridgeway*, Oxford: Oxford University School of Archaeology monograph series

McOmish, D (1996) 'East Chisenbury: ritual and rubbish at the British Bronze Age-Iron Age transition' 70 *Antiquity* 68–76

Mack, J (1986) *Madagascar: Island of the Ancestors*, London: British Museum

Macphail, RI (1996) 'East Chisenbury, Wiltshire: brief assessment of soil micromorphology', unpublished report, Institute of Archaeology, University College London

Moore, J (1997) 'The infernal cycle of fire ecology', in Topping, P (ed), *Neolithic Landscapes*, Oxford: Oxbow

Owoc, MA (2000) *Aspects of Ceremonial Burial in the Bronze Age of South-west Britain*, unpublished PhD thesis, University of Sheffield

Owoc, MA (2001) 'The times, they are a changin': experiencing continuity and development in the Early Bronze Age funerary rituals of south-western Britain', in Brück, J (ed), *Bronze Age Landscapes: Tradition and Transformation*, Oxford: Oxbow

Parker Pearson, M (1995) 'Southwestern Bronze Age pottery', in Kinnes, I and Varndell, G (eds), *'Unbaked Urns of Rudely Shape': Essays on British and Irish Pottery for Ian Longworth*, Oxford: Oxbow

Parker Pearson, M (1999) *The Archaeology of Death and Burial*, Stroud: Sutton

Parker Pearson, M and Ramilisonina (1998) 'Stonehenge for the ancestors: the stones pass on the message' 72 *Antiquity* 308–26

Parker Pearson, M and Sharples, N with Mulville, J and Smith, H (1999) *Between Land and Sea: Excavations at Dun Vulan, South Uist*, SEARCH volume 4, Sheffield: Sheffield Academic Press

Parker Pearson, M, Richards, C, Allen, M, Payne, A and Welham, K (2004) 'The Stonehenge Riverside project: research design and initial results' 14 *Journal of Nordic Archaeological Science* 45–60

Parker Pearson, M, Sharples, N and Symonds, J with Mulville, J, Raven, J, Smith, H and Woolf, A (2004) *South Uist: Archaeology and History of a Hebridean Island*, Stroud: Tempus

Passmore, AD (1940) 'A disc barrow containing curious flints near Stonehenge' 49 *Wiltshire Archaeological and Natural History Magazine* 238

Pryor, F (2003) *Britain BC: Life in Britain and Ireland Before the Romans*, London: HarperCollins

Richards, C (1996) 'Henges and water: towards an elemental understanding of monumentality and landscape in Late Neolithic Britain' 1 *Journal of Material Culture Studies* 313–36

Rival, L (ed) (1998) *The Social Life of Trees*, Oxford: Berg

Selkirk, A (2004) 'Gayhurst barrow cemetery' 16(11) *Current Archaeology* 510–11

Stone, JFS (1935a) 'Some discoveries at Ratfyn, Amesbury and their bearing on the date of Woodhenge' 47 *Wiltshire Archaeological and Natural History Magazine* 55–67

Stone, JFS (1935b) 'Excavations at Easton Down, Winterslow, 1933–1934' 47 *Wiltshire Archaeological and Natural History Magazine* 68–80

Stone, JFS (1949) 'Some Grooved Ware pottery from the Woodhenge area' 15 *Proceedings of the Prehistoric Society* 122–27

Stone, JFS and Young, WEV (1948) 'Two pits of Grooved Ware date near Woodhenge' 52 *Wiltshire Archaeological and Natural History Magazine* 287–306

Strathern, A (1971) *Self-decoration in Mount Hagen*, London: Duckworth

Thomas, J (1996) *Time, Culture and Identity: An Interpretive Archaeology*, London: Routledge

Thomas, J (1997) 'An economy of substances in Early Neolithic Britain', in Robb, J (ed), *Material Symbols: Culture and Economy in Prehistory*, Carbondale: Southern Illinois University Press

Tilley, C (1999) *Metaphor and Material Culture*, Oxford: Blackwell

Tilley, C (2004) *The Materiality of Stone*, London: Berg

Vatcher, F de M (1969) 'Two incised chalk plaques near Stonehenge Bottom' 43 *Antiquity* 310–11

Wainwright, GJ with Longworth, IH (1971) *Durrington Walls: Excavations 1966–1968*, London: Society of Antiquaries

Whittle, AWR (1997) 'Remembered and imagined belongings: Stonehenge in its traditions and structures of meaning', in Renfrew, C and Cunliffe, B (eds), *Science and Stonehenge*, Proceedings of the British Academy 92, Oxford: Oxford University Press

Woodward, AB (2000) *British Barrows: A Matter of Life and Death*, Stroud: Tempus

Woodward, AB and Woodward, PJ (1996) 'The topography of some barrow cemeteries in Bronze Age Wessex' 62 *Proceedings of the Prehistoric Society* 275–91

– 5 –

THE MIRROR OF THE SUN

SURFACE, MINERAL APPLICATIONS AND INTERFACE IN CALIFORNIA ROCK-ART

David Robinson

INTRODUCTION

Along the Vandenberg coastal plain of California, Beehive Shelter overlooks the eastern shores of the Pacific Ocean (**Figure 5.1 colour section**). In areas where survival is possible, faint black geometric lines and amorphous splodges of red pigment can be seen smeared on the surface of the shelter wall. In comparison to the vibrant and visually compelling polychrome pictographs typifying rock-art produced by the coastal indigenous peoples known as the Chumash, the pigment at Beehive Shelter hardly merits attention. There are no anthropomorphic or zoomorphic creatures to be seen, no series of deftly applied dotted or fine lined applications, nor any of the elaborate discs that typify Chumash rock-art (Grant 1965). At nearby Rattlesnake Shelter, similar non-representational swaths of red pigment also defy conventional categorisation (**Figure 5.2 colour section**). These red smears cannot comfortably be interpreted as representational images of animals, plants, tools, or entoptic phenomena. The marks seem to stand in the simplest of terms for what they are – red pigment on rock.

Much is well understood of the mineral constituents of these and other pigments forming the pictographs of the native Californian Chumash. Samples taken from exfoliated fragments provide scientific information: haematite, an iron oxide earth, is utilised for red pigment; manganese is often used for black; lime for white; mineral earth for greens (see Grant 1965; Scott and Hyder 1993; Scott *et al* 2002). These studies have the effect of bringing into a scientific frame of reference the material constituents that make up rock-art. We understand rock-art as the product of quarrying, mixing, and application processes, all easily recognisable activities from our modern vantage point. What these studies do not do is provide an understanding of the attitudes or beliefs that the users of these minerals held that led to the kinds of practices observed at Beehive and Rattlesnake Shelters.

While the type of knowledge we gain through scientific analysis *is* vital knowledge to understanding past practices, such analysis operates by translating those indigenous practices – in this case the making of rock-art – into modern Western scientific terms. This chapter attempts to complement these scientific understandings, by drawing upon a set of generally tangential discourses from Native American philosophy, ethnography and archaeology. The point here is not to make understandings of indigenous practices conform to our Western view, nor is the purpose to jettison our modern pre-understandings and enter a wholly other discursive realm. Rather, the aim is to fuse relevant information from a variety of avenues so as to achieve better interpretations of indigenous engagements with the mineral world. In this manner, an explanation of the often overlooked and little discussed non-representational smears of mineral pigment at Chumash sites in the Vandenberg coastal region (**Figure 5.3**) can be more readily approached.[1] It is further suggested that a better understanding of the non-representational images can also contribute to a more informed interpretation of some of the representational rock-art in the region.

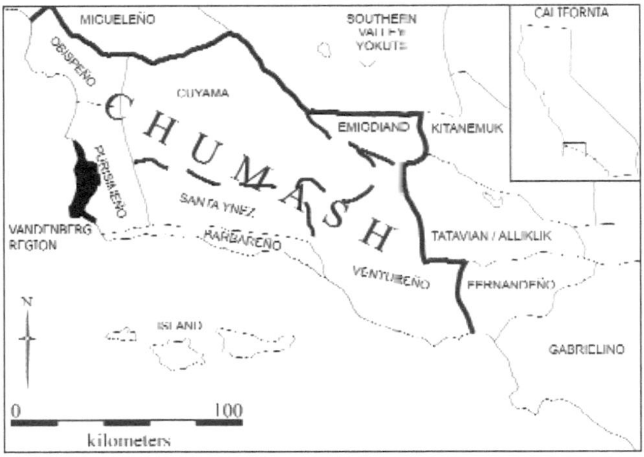

Figure 5.3: Map of the Chumash and neighbouring linguistic divisions. Rock-art sites discussed in this paper are in the Vandenberg region, which is shown in black (map created by Ian Forbes).

BREAKING TEETH: NORTH AMERICAN INDIGENOUS PHILOSOPHY

From the perspective of mainstream North American society, *native* is a word invariably analogous with 'other'. The *ancestral* traditions of that other are the interest of the archaeologist, who attempts to understand this other-in-the-past through the study of material culture surviving to the present. Julian Thomas has suggested that this very otherness that the past embodies is in large part what attracts us to study it; however, 'prehistoric archaeology systematically erases those qualities of the past' in the process of making it familiar (Thomas

1999: 1). There is an uneasy dialectic in effect when attempting to balance between the construction of interpretation and the preservation of otherness.

The survival of non-representational rock-art in the form of red pigment, such as that described at Beehive Shelter, epitomises this other-in-the-past from our view in the present. However, having no form or imagery to be admired, the pigment smears do not capture academic attention, slipping through rock-art analyses almost unnoticed. It is primarily because of laws dealing with cultural resource management that this rock-art is even documented at all. Just as the formless marks, being 'other', are rarely looked upon, alternate modes of reasoning offered by indigenous voices are rarely listened to. Native American philosophy is overlooked in part because, like the paint smudges, it is foreign to mainstream understanding. Indigenous writers fully recognise this condition, and point out that mainstream analytical approaches (academic and otherwise) fail to appreciate the fundamental differences between tribal and 'Western' cultures because the assumptions underlying these approaches originate external to their object of study: 'Tribal society is of such a nature that one must experience it from the inside. It is holistic, and logical analysis will only return you to your starting premise none the wiser for the trip' (Deloria 1970: 12). Western analyses have the effect of consuming the other while failing to understand it from its own terms. From the Native American viewpoint, this failure only accentuates the fact that indigenous North American belief systems embody 'a *different way of Being*' (Highwater 1981, emphasis in original; see also Deloria 1970: 85). It is because of this difference that linear Western processes of logic are said to fail when confronted by alternate indigenous modes of thought. As Antonia Machado-Ruiz states, 'the other refuses to disappear; it subsists, it persists; it is the hard bone on which reason breaks its teeth' (in Highwater 1981: 11). The persistence of the smears of red pigment that defy attempts at orderly classification and interpretation in rock-art studies illustrates this failure to succumb to mainstream Western modes of understanding.

Fortunately, Native American philosophers actively present avenues of dialogue to reach out to their 'other'. Continuing with Deloria:

> The best method of communicating Indian values is to find points at which issues appear to be related. Because tribal society is integrated toward a centre and non-Indian society is oriented toward linear development, the process might be compared to describing a circle surrounded with tangent lines. The points at which the lines touch the circumference of the circle are the issues and ideas that can be shared by Indians and other groups. There are a great many points at which tangents occur, and they may be considered as windows through which Indians and non-Indians can glimpse each other. Once this structural device is used and understood, non-Indians, using a tribal point of view, can better understand themselves and their relationship to Indian people. (1970: 12)

In essence, this description of circular intersection harmonises quite well with theories of hermeneutics, particularly Gadamer's (1975) rather two-dimensional image of the 'fusion of horizons'. It is therefore possible that tangents between

Western discourse and indigenous philosophy can provide avenues of understanding from which the theoretical underpinnings of archaeological interpretation can be derived. This chapter does not, however, argue that indigenous philosophies are static entities, unchanged over the millennia. While the relationship between contemporary Native American groups and those whose remains are studied by prehistoric archaeologists in North America is acknowledged, Native American philosophy is primarily used here as a foil to the Western philosophies that more usually inform archaeological interpretation, whether processual or post-processual. Through the incorporation of indigenous philosophy within a hermeneutic exercise, an interpretation of the non-representational pigment applied at California rock-art sites like Beehive Shelter can be attempted.

SENTIENCE EXPRESSED IN MOTION

Since the rock-art discussed here does not have any recognisable form, analysis must move away from an image-based consideration by foregrounding the process of pigment application itself. At a fundamental level, the application of pigment to rock involves bodily movements and interactions with the rock surface. Speaking from a Native American point of view, the philosopher Highwater considers bodily movement to be 'humanity's most fundamental and expressive act' (1981: 137). In discussing Native American dance and performance, he argues that they constitute a 'direct, nonverbal, unreasoned assertion of ideas and sentience expressed in forms of motion' (*ibid*). What is of importance here is the *physicality of action* in many aspects of Native American practice. Rather than focusing upon the mind of the performer, an essential aspect of indigenous performance is the unfolding of meaning within and through the actions of an embodied sentient self. Sentience is also important in that as a term encompassing all aspects of being, it foregrounds humans (and other beings) as agents outwardly active in a world of substance. The activities of a being collapse cognition and action into instantaneous expression. Sentience and its expression are key to circumventing the modernist trap of mind/body duality, and the concept permeates multiple facets of indigenous practice. Highwater posits similar views with respect to modern native literature: 'The *dramatis personae* of tribal literature is essentially external. The reader tends to know characters by their actions and not by an outpouring of feelings and various states of mind' (*ibid*: 117). Mythologist Karl Kroeber (a non-Native) echoes this sentiment in considering oral mythology. He terms mythic tellings 'enactments' or narratives of 'action-behaviour' (1998: 90). Character and motivation are revealed through *actions*. Furthermore, the telling of a myth is a social happening, heard and seen by a group: '... an enactment therefore is determined by its audience's active participation' in conjunction with the teller (*ibid*: 126). Thus the basic postulate expressed here is that action in the real world both reveals and takes precedence over internal dialogues.

Returning again to the amorphous pigment smudges at Beehive and Rattlesnake Shelters, we begin, in light of the foregoing discussion of Native American philosophy and myth, to see the basis for some possible insights into the enigmatic marks. Viewed from the perspective of the Native American conception described above, the smudges of pigment emerge as possible evidence of expressive enactments occurring on site, involving an active interplay between the mediums of pigment and rock surface. These acts of application can be suggested to have been part of a set of 'action-behaviours' with the painting as simply the remaining archaeological evidence. Importantly, though, the on-the-rock pigment would also be an enduring reminder, memorialising the individual enactment and the cultural processes behind its creation. If this event was public, then it follows that the bodily movements resulting in the creation of the smears of rock-art involved visual communication between the painter and watchers; if the act was private, the personal importance of this rock-art would reside more in the tactile encounter between the organ of skin and the rock-surface through the facilitation of pigment and binding agent. This tactile facet is suggested at another recently discovered Vandenberg rock-art site (CA-SBA-612). Within the shelter, pigment was applied to a fossilised shell contained within the rock wall. This pigment appears to have been applied by finger, and it is possible that the smear was placed on the shell during an episode of body painting (Dan Reeves, pers comm). Indeed, it is likely that body painting accompanied the making of rock-art on at least some occasions, as the ethnographic record overwhelmingly suggests that the most prevalent use of mineral pigments was on the body during public performances such as dances and other ceremonies (see for instance Gayton 1948a, 1948b; Gifford 1955; Librado 1977). As will be discussed throughout the remainder of this chapter, the materiality of pigment offered many avenues for expression and the interlinking of various kinds of sentience; at the most personal level, however, its tactile quality afforded people a non-verbal, on-the-skin medium of expression. Thus pigment was bound within a wide tradition of bodily movement, shared performance, and unfolding enactments, with rock-art as one aspect of this ongoing expression of sentience.

ETHNOGRAPHIC HORIZON: MINERAL, BODY AND POWER

The ethnographic record can further illuminate the interplay between pigment, body and power amongst the Chumash. Another Vandenberg rock-art site featuring extensive use of red mineral pigment is Honda Ridge, where the rock-art is characterised by numerous red linear smears and a few possible representational and/or geometric shapes. Hyder has classified Honda as a puberty-initiation site based upon comparisons with sites from other regions (see Hyder *et al* 1996). At Honda, fingerprints can still be seen in some of the smears (**Figure 5.4 colour section**), confirming again that skin was employed to produce at least a portion of the rock-art. If, as suggested above, the action of application itself was primary in importance, puberty sites can be thought of as places where young adults negotiated the transition to adulthood through, in part, an introduction to the culturally important practice of applying pigment to rock. This established a personal, tactile relationship with pigments at the first stage of adulthood. That ochre was used on the skin in some puberty rites in California is certain: a Kitanemuk woman recounted being 'rubbed all over with red ochre' by her mother after her first menstruation – the pigment remained for four months as she was prohibited from bathing as part of the rite (Hudson and Blackburn 1986: 45–46). Evidence of ochre smeared on the body in Chumash burials (see for instance King 1982) provides another example of the way in which important transitional moments within the adult life cycle featured an engagement between the body and the materiality of pigments.

Though ethnographers rarely asked about the significance of body paint beyond inquiring into the meaning of design patterns, there is some suggestion in the ethnographic literature that certain minerals were considered to possess special properties that had implications for their use in body painting. Interestingly, these properties appear to have been more than simply symbolic, in the Western sense of the term (eg Kroeber 1907; Mayfield 1993: 83), and to have been closely linked to indigenous concepts of power and its operation. For example, Bob Batista, a famous Southern Yokuts doctor, 'combined burned Jimson seeds with a small amount of *'ajap'* [a white mineral substance] mixed with water to make his body paint for dances' (Gordon 1996: 45). *'Ajap'* was considered to possess supernatural power, while jimson weed is a hallucinogenic plant (personified by the Chumash as a powerful grandmother entity). Jimson weed was an ingested visionary substance which was considered to be imbued with *'atiswin*. *'Atiswin* is a complex concept having no analogue in English: it seems to stand for both supernatural beings and the *quality* of power they embody. It also stands for powerful substances, items or talismans containing powers, and implies the ability of a person to utilise those powers (see Applegate 1978 and Blackburn 1975). A person with access to numerous talismans or the ability to draw upon supernatural power was considered to be 'very *'atiswinic'*. Thus we can see that in some cases, pigment itself would have been imbued with *'atiswin* and its usage would have involved the deployment of its power.

PLACES: QUARRIES, MYTH AND SUBSTANCES

For the Chumash, many things could potentially contain *'atiswin*, including places. The environmental or geological attributes specific to a locality lent to it its own unique quality or power. However, each place also had its own history of usage, with individuals interacting and perceiving it in different ways. As Vine Deloria suggests:

> The places where revelations were experienced were remembered and set aside as locations where, through rituals and ceremonials, the people could once again communicate with the spirits. ... It was not what people believed

> to be true that was important but what they experienced as true. Hence revelation was seen as a continuous process of adjustment to the natural surroundings and not as a specific message valid for all times and places. (1994: 66–67)

In this sense, the power *experienced* at a place was deeply implicated in cultural attitudes towards environments and geologies. Activities undertaken engaged people with and confirmed the presence of *'atiswin* inherent at a locality. Importantly to the discussion here, substances removed could also retain that *'atiswin*.

In a wider California context, many different types of quarry sites and the minerals they contained were, at times, considered to be places that contained *'atiswin*. Ethnographic information on obsidian procurement, for example, indicates that a Wintu trek to an obsidian source was a 'semireligious quest' entailing fasting. For many groups, obsidian itself reputedly had toxic qualities (see Heizer and Treganza 1944: 303, 305; Mayfield 1993: 82–83), and proper rules of behaviour were necessary to interact with the power present at sites where obsidian was obtained or used. For the Maidu, improper behaviour would lead to failure in flint knapping or hunting kills (Heizer and Treganza 1944: 331). In Chumash territory, excavations at a microblade quarry revealed a cache of abalone shells and an upper raven bill, and it has been noted that 'these unusual items at this remote quarry location suggest the possibility of ritual behaviour' (Arnold 1987: 140). Closer to Vandenberg, oral history records that during the Mission Period, the process of quarrying minerals involved prescribed behaviour with negative consequences for transgressors:

> At a place a short distance south of Santa Inez Mission there is a bank where they once dug lime for the building of the mission ... there were two island men digging lime there at the top of the bank, and the others told them not to talk too loudly or something might happen to them, but they paid no attention. Suddenly they heard a sound like an explosion and everyone looked up. They saw a cloud and in the cloud was an animal that looked like a serpent. It had a big head with a tuft on it like a quail ... the two islanders were carried off by the cloud ... (Blackburn 1975: 299)

The use of minerals created interconnections between the place of extraction, the place of use, and the social setting of usage. Gifford recorded that in the Akantoto dance of the central Miwok, a dancer 'rubbed himself all over with powdered human legbones, to make him immune from the heat in the upper part of the ceremonial house, where he mostly performs. The powdered bones are obtained from a hole ... where the rock giant Yayali is said to have thrown the bones of people he devoured. These bones are also used for medicine' (1955: 274). In this instance, the conceptualisation of a white mineral as legbone presented a physical and tactile link between the body and the local mythic landscape. Indeed, as evidenced in oral narratives, distinctively shaped rock formations were sometimes explained as places where mythic events occurred: mythic beings were sometimes turned to stone, thus forming the rock (see Latta 1936 and Kroeber 1907). Circular rock formations at places considered culturally important were in some instances interpreted as circles of mythic dancers turned to stone, with one such case explaining the rock-art as the face painting of those ossified mythological dancers (see Cummins 1992). Indeed, as already stated, the special qualities of particular places could be at least partially based upon the *properties* of the geological matrix (as exemplified by the Miwok white mineral). In particular, shiny substances like obsidian and quartz crystal were often interpreted from mythological understandings of the landscape (see eg Heizer and Treganza 1944; Blackburn 1975). It is clear that minerals were bound within a nexus of interrelationships involving the body, concepts of power (ie *'atiswin*), mythological understandings and experiences of places, and reflective substances. In turning now to the Vandenberg region, it will be seen that the quality of reflection was of particular interest to its inhabitants.

MIRROR OF THE SUN

Just as the geology of a mineral extraction site and the material derived from it were filled with cultural meaning, the geology of the rock-art sites themselves was also likely considered important by those responsible for the creation of the pictographs we see today. Crucially, the rock surface or its inclusions at all four Vandenberg sites mentioned thus far have reflective qualities. The rock formations at the Beehive and Rattlesnake Shelters have crystalline bands folded throughout their matrices, and at Rattlesnake in particular, abundant direct sunlight reflects off the crystals. The fossilised shell at SBA-612 likewise has an iridescent quality (Rick Bury, pers comm). However, Honda Ridge is the Vandenberg rock-art site that is most striking in its reflective properties (**Figure 5.5 colour section**).

At Honda Ridge, the surface of the rock was formed under the earth by a process of joint movement caused by the tectonic activity in the region. The uplift of the plane of this rock exposed the surface, called a slickenside, to the open air. The surface is smooth, inviting to the touch, and at places so shiny that your reflection is mirrored back to you.

The Chumash name for the earth was Hutash, or 'the mirror of the Sun'.[2] The famous Chumash consultant, Fernando Librado, who worked as a *vaquero* (ranch hand) in the region in the late 1800s, reported what his grandfather had taught: 'All mankind lives in the sphere of Hutash. The sphere of Hutash is everything that contains the human race, and for the Indians the sphere of Hutash is the land. All are born of the earth, and the sun revivifies all creatures ...' (Librado 1977: 52–53). Rituals suggest a reciprocal relationship between the Sun and Hutash. It appears as if the Chumash conceptualised themselves as intermediaries within this relationship. For instance, winter solstice group ceremonies climaxed with a collective urging of the Sun to return north from its winter position on the horizon and encourage the eventual onset of spring (*ibid*). However, the human position was characterised by ambiguity and the realisation that the Sun was the ultimate authority. One ceremony included the words: 'None of us in this assembly control our destiny, for we live in the shadow of the Sun ... Hutash is the mirror of the Sun, and the Sun is the mirror of Hutash' (*ibid*). At Honda Ridge, Hutash literally is the mirror of the sun.[3] As places with unique reflective qualities, these sites and the pigment they contain are situated exactly at the most visible point of that relationship. A person applying pigment would be interacting with both sun and earth, as well as drawing upon and entering into associations related to the pigment itself.

MULTIPLE AND EXTENDED INTERFACES

Just as the application of non-representational rock-art allowed practitioners to interface with powers such as the Sun and Hutash, I wish to suggest that representational imagery (ie geometric and zoomorphic shapes) similarly involved concepts of interface within which the artist could participate. It is important to realise in this regard that for some California groups, the *creation* of designs was a way to *acknowledge* and *activate* power. In the creation of geometric designs, the Wintu, for example, maintain 'a concept of the immutability of essence and transiency of form' (Lee 1975: 133). Form itself is 'shifting, and relatively unimportant' and Lee notes that: 'Matter and relationships, essence, quality are all given. The Wintu actualises a design endowing it with temporality and form through his experience. But he neither creates nor changes; the design remains immutable' (*ibid*: 131). Contemporary Native American art shares a similar orientation. Highwater, for example, speaking of Native American art objects, notes that '[t]he objects of the Indians are expressive and not decorative because they are alive, living in our experience of them' (Highwater 1981: 77). In the case of pictographs, the materiality of pigment enabled the interface between pre-existing form and artist to become actualised. Thus pigment itself, as the agent of actualisation, was a material with interface valences of its own. As the example of mixing *'ajap'* with jimson weed to make paint suggests, mineral substances were important not only in and of themselves, but in particular through their alchemic applications. As the interface of the rock surface engaged the practitioner with the power of the place and provided the medium for the actualisation of design, so mineral pigment and its admixture extended that interface by multiplying associations according to its ingredients. For instance, pigment containing jimson weed may have been considered as having special potency related to the power attributed to the plant. The painting of geometric form can be thought of as a polyvalent act multiplying allusions to the powers specific to place, attributes accorded to the pigment, and the actualisation of design, for example.

As will be demonstrated below, in the creation of zoomorphic rock-art, this interface was further extended to command the agency of the sentient beings depicted. Not only the sun and jimson weed, but insects, animals and creatures of the sea were the most immediate and prevalent source of power in Chumash land and seascapes. While rare in the Vandenberg region, zoomorphic images are frequent in the wider realm of Chumash rock-art. It is therefore possible that, like non-representational rock-art, zoomorphic pictographs sometimes incorporated aspects of enactment in the creation of imagery with pigment admixture playing a role; however, the added dimension of corresponding form alludes to *specific* relationships and their activation.

Some zoomorphic representational imagery does exist in the Vandenberg region: the aptly named Swordfish Cave contains a prominent pictograph of a swordfish. Swordfish were considered 'people' of the ocean and it is clear that they were believed to be sentient, powerful beings (for the importance of swordfish to the Chumash, see Davenport *et al* 1993). However, the perspective I wish to emphasise here is how this representational painting implies some kind of interface to or concern with the aquatic environment of the ocean – an environment removed from the site itself. Swordfish Cave is inland, separated from the coast by the formidable Tranquillon ridge. Water seeping from the back of the cave perhaps implies a linking to the waters along the coast via a conceptual underground, watery passage (see discussion in Whitley 2000: 122–23). Within Swordfish Cave, there is furthermore to be found an etching on the rock surface spelling out the words 'La Puerta', Spanish for 'The Door' (**Figure 5.6**). It has been speculated that perhaps the Chumash consultant Fernando Librado incised these words, since he worked as a *vaquero* in the region (see Hyder *et al* 1996). The etching of the words 'La Puerta' offers further indication that Swordfish Cave has indeed (at least in historic times) been conceptualised as a passageway, or opening. As a kind of doorway, it perhaps presented a far-reaching interface linking the practitioner with remote powers of agency.

Another example of how the use of pigment enabled interaction with distant powers across aquatic environments is found in a native account of sorcery on Santa Catalina Island (see Hudson 1979). In 1801, an outbreak of sickness on the mainland was attributed to three sorcerers on the island:

Figure 5.6 The words 'La Puerta' incised within Swordfish Cave (© Rick Bury).

> There in the canyon, where none save the three could see it, they painted the world on the level ground. They painted it like the world is, and they painted all, outdoors in the canyon ... When they were finished painting the world, they began to paint infirmities and blood, as well as such dangerous animals that cause pain by biting, clawing, and so on ... all of these things for sickness ... Meanwhile the people in San Gabriel were dying. They had headaches, and blood came out of the mouth ... (*ibid*: 357)

In response to this situation, an expedition was launched to the island, the sorcerers killed and the painting destroyed, thus ending the epidemic (*ibid*: 358). This account clearly demonstrates that in coastal California, the use of pigment and the creation of zoomorphic designs was considered to be able to influence events in remote localities by activating the powers that sentient beings possess.

CONCLUSION

As the foregoing discussion has illustrated, the employment of Native American philosophy and ethnography can aid the archaeological interpretation of non-representational rock-art. In particular, the Native American understanding of enactment and the relationship between sentience and motion as played out through practices of dance and myth seem to correlate to a parallel understanding with respect to performance and the application of pigment to rock surface. In this interpretation, mineral pigment and its physical, performative application facilitates connections between people and other multiple sentient powers existing in the immediate and extended environment. The pigment and its inclusions reference their site of extraction and may contain special transportable qualities. Far from being inert, mineral substances within California indigenous terms were redolent with potency. The unique geology of place at a rock-art site referenced power to be engaged with; the smearing of mineral pigment was a cultural enactment linking the individual at various stages of adulthood with indigenous philosophical postulates and, in the case of the Chumash, concepts of *'atiswin*.

As this volume demonstrates, the study of mineral substances and past understandings of them entangles archaeologists in a variety of facets of past life. Indeed the mineral topic provides an interpretative fulcrum from which to pivot into disparate aspects of past behaviour. The interrelated nature of these aspects means that our understanding of past conceptions of the material realm, including the mineral world, is enhanced when we fuse together a variety of related discursive elements. This chapter has therefore addressed a range of topics, from Native American performance to ethnohistorical discussions of mineral quarrying. Given that the application of mineral pigments to rock surfaces by the Chumash drew upon a multiplicity of relationships, it is both appropriate and necessary to explore these relationships by recourse to diverse sources of information, including ethnography, archaeology, and the philosophy of modern indigenous writers. Through engagement in this kind of hermeneutic exercise, it becomes possible to retain a sense of otherness while constructing an archaeological interpretation. We should do more than simply project a mirror image of modernity upon the past.

Returning to Rattlesnake Shelter and the issue of the persistence of the pigment on the rock through time, I would like to finish by discussing the final episode in the creation of rock-art at the site. Like the words 'La Puerta' at Swordfish Cave, this final image was incised, and consists of a rendering of a European ship that has been etched into the underlying red pigment (**Figure 5.7**). In an illustration of the close relationship between rock-art and the experience of power, the etching references the influx of the Spanish *entrada* and the incursion of a new force into the complex realm of power and sentience that was the Chumash world. It is significant that the pre-existing pigment provided a medium for subsequent engagement. The etching into the earlier coat of red mineral pigment is evidence that into the historic period, the Chumash continued to employ and at the same time modify traditional strategies to negotiate the fluctuating power relations that characterise the sphere of Hutash.

Figure 5.7: Rendering of a masted vessel incised into the red pigment of Rattlesnake Shelter (© Rick Bury).

Notes

1 The inhabitants of the Vandenberg region were part of the linguistically defined Chumash subgroup called the Purisemeño. The Purisemeño and their predecessors were maritime coastal hunter-gatherers occupying the region from approximately 9000 BP into the late 1700s AD (see Glassow 1996).

2 The term *Hutash* is a case in point for the difficulty presented to modernist analyses. *Hutash* was a polysemous term, sometimes meaning the earth, sometimes referring to the planet Venus (see Blackburn 1975), and sometimes meaning *chia*, an important seed-food for the Chumash (see Timbrook *et al* 1993). It is difficult to ascertain whether the multiple meanings of the word indicate an ontologically differing viewpoint on categorisation or simply metaphorical associations.

3 This interface between sun and earth is reiterated at another Vandenberg rock-art site called Window Cave – a small rock shelter with a natural 'window' in the side allowing in a shaft of light at the time of Winter Solstice sunset. The shaft of light reaches a natural crack – next to the crack is a spoked-symbol connecting to other incised petroglyphs (see Spanne 1985). Spanne has argued that this rock-art site can be thought of as an observatory. From the perspective of this chapter, the window and possibly the crack provide both a boundary and an *interface* between sun and earth in the confines of the shelter.

Acknowledgments

I would like to thank Rick Bury for his generosity in allowing me to use his excellent photographs as well as for his expertise in the field. I would also like to acknowledge him along with Carole Bury, Christopher Chippindale, Adam Brumm and Nicky Boivin for their comments on the initial drafts of this paper. Special thanks go to Nicky Boivin and Mary Ann Owoc for the invitation to participate in this volume. Thanks also to Larry Spanne, Bob Peterson and the staff at Vandenberg as well as Mike Glassow, John Johnson, Antoinette Padget, Dan Reeves and Bonnie Whitley.

References

Applegate, R (1978) ''*Atishwin*: the dream helper in South-Central California' 13 *Ballena Press Anthropological Papers*

Arnold, JE (1987) *Craft Specialization in the Prehistoric Channel Islands, California*, Berkeley: University of California Press

Blackburn, TC (1975) *December's Child: A Book of Chumash Oral Narratives*, Berkeley: University of California Press

Cummins, MW (1992) *How Coyote Stole the Sun: The Myth, the Music, and Other Features of the Yokuts Culture*, Fresno: Braun-Brumfield

Davenport, D, Johnson, JR and Timbrook, J (1993) 'The Chumash and the swordfish' 67(255) *Antiquity* 257

Deloria Jr, V (1970) *We Talk, You Listen: New Tribes, New Turf*, New York: Macmillan

Deloria Jr, V (1994) *God is Red: A Native View of Religion*, Golden: Fulcrum

Gadamer, HG (1975) *Truth and Method*, London: Sheen and Ward

Gayton, AH (1948a) 'Yokuts and Western mono ethnography. 1: Tulare Lake, Southern Valley, and Central Foothill Yokuts' 10(1) *Anthropological Records*, Berkeley and Los Angeles: University of California Press

Gayton, AH (1948b) 'Yokuts and Western mono ethnography. II: Northern Foothill Yokuts and Western Mono' 10(2) *Anthropological Records*, Berkeley and Los Angeles: University of California Press

Gifford, EW (1955) 'Central Miwok ceremonies' 14(4) *Anthropological Records*, Berkeley and Los Angeles: University of California Press

Glassow, M (1996) *Purisemeño Chumash Prehistory: Maritime Adaptations Along the Southern California Coast*, Fort Worth: Harcourt Brace

Gordon, M (1996) 'An ethnographic compilation of the sources, composition, and uses of paints by the Yokuts of the Southern San Joaquin Valley and Sierra Nevada, California' 7 *Kern County Archaeological Society Journal* 36

Grant, C (1965) *The Rock Paintings of the Chumash*, Berkeley: University of California Press

Heizer, RF and Treganza, AE (1944) 'Mines and quarries of the Indians of California' 40(3) *California Journal of Mines and Geology* 291

Highwater, J (1981) *The Primal Mind: Vision and Reality in Indian America*, New York: Meridian

Hudson, T (1979) 'A rare account of Gabrielino shamanism from the notes of John P Harrington' 1(2) *Journal of California and Great Basin Anthropology* 356

Hudson, T and Blackburn, TC (1986) 'The material culture of the Chumash interaction sphere, Volume IV: ceremonial paraphernalia, games, and amusements' 30 *Ballena Press Anthropological Papers*

Hyder, WD, Lee, G and Hogue, K (1996) 'Final technical report: six rock art sites at Vandenberg Air Force Base, Santa Barbara County, California', unpublished technical report filed with Science Applications International Corporation, Santa Barbara, California

King, LB (1982) *Medea Creek Cemetery: Late Inland Chumash Patterns of Social Organization, Exchange and Warfare*, unpublished PhD thesis, University of California, Los Angeles

Kroeber, AL (1907) 'Indian myths of South Central California' *University of California Publications: American Archaeology and Ethnology* 4(4), Berkeley, CA: University of California Press

Kroeber, K (1998) *Artistry in Native American Myths*, Lincoln: University of Nebraska Press

Latta, F (1936) *California Indian Folklore*, Santa Cruz: Bear State

Lee, D (1975) 'Linguistic reflections of Wintu thought', in Tedlock, D and Tedlock, B (eds), *Teachings from the American Earth: Indian Religion and Philosophy*, New York: Liveright

Librado, F (1977) *The Eye of the Flute*, Banning: Malki Museum Press

Mayfield, TJ (1993) *Indian Summer: Traditional Life among the Choinumne Indians of California's San Joaquin Valley*, Berkeley, CA: Heyday Books and California Historical Society

Scott, DA and Hyder, WD (1993) 'A study of some Californian Indian rock art pigments' 38(3) *Studies in Conservation* 155

Scott, DA, Scheerer, S and Reeves, DJ (2002) 'Technical examination of some rock art pigments and encrustations from the Chumash Indian site of San Emigdio, California' 47(3) *Studies in Conservation* 184

Spanne, LW (1985) 'Rock art sites as solar observatories: two possible examples from Vandenberg Air Force Base', in Benson, A and Hoskinson, T (eds), *Earth and Sky*, Thousand Oaks, CA: Slo'w Press

Thomas, J (1999) *Understanding the Neolithic*, London: Routledge

Timbrook, J, Johnson, JR and Earle, DE (1993) 'Vegetation burning by the Chumash', in Blackburn, TC and Anderson, K (eds), *Before the Wilderness: Environmental Management by Native Californians*, Menlo Park: Ballena Press

Whitley, DS (2000) *Art of the Shaman: Rock Art of California*, Salt Lake City: University of Utah Press

– 6 –

A PHENOMENOLOGY OF THE BURIED LANDSCAPE

SOIL AS MATERIAL CULTURE IN THE BRONZE AGE OF SOUTH-WEST BRITAIN

Mary Ann Owoc

INTRODUCTION

In the following contribution, a discussion of late Neolithic and early Bronze Age funerary/ritual sites in south-west Britain focuses upon their builders' use of stone and soil. It is suggested that these elements were meaningfully constructed and, moreover, used spatially and temporally in combination with other items of material culture to create and reproduce particular local traditions of knowledge. A number of examples and a case study are presented. The *community* nature of these monuments is stressed within a discussion of the appropriate application of phenomenology to archaeological accounts addressing the relationships between humans and their surface and buried landscapes.

THE LITHOSPHERE AS MATERIAL CULTURE

In the last 30 years or so archaeologists have greatly expanded the traditional definitions of material culture to include things not normally considered to be artefactual (Schlereth 1969). In attempting to illustrate the importance of 'material culture' to an increasingly self-aware discipline, James Deetz (1977: 357) defined the term as 'that sector of our physical environment that we modify through culturally determined behavior'. Under this broad definition, things like water (as pond), soil (as garden), and stone (as wall) are all capable of operating through human agency to enable us to 'cope with the ... world, facilitate social intercourse, delight our fancy, and create symbols of meaning' (Herskovitz 1963: 119). Lithospheric elements, then, once recognised, named or used, become capable of operating in signifying systems that facilitate knowledgable encounters with the world. In an effort to contribute to this broader understanding of material culture, and the place of the mineral world within it, this contribution addresses a series of deliberate, and highly public, engagements with the mineral world occurring in south-west Britain between c 2500 and 1500 cal BC (understood as the later Neolithic and earlier Bronze Age) (**Figure 6.1**).

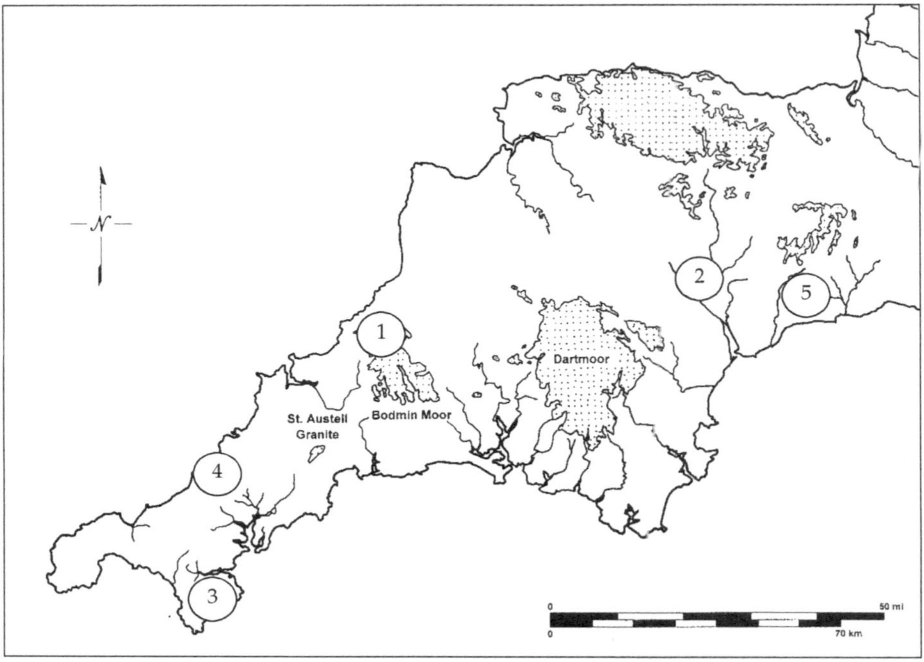

Figure 6.1: South-west Britain noting sites or regions detailed in the text: (1) Davidstow Moor Cemetery; (2) Upton Pyne; (3) Trelen 2; (4) Crig-a-Mennis; (5) Broad Down-Farway cemetery.

THE WOOD FOR THE TREES?

The most visible remains of the later Neolithic/earlier Bronze Age today are the earthen mounds, rings, stone cairns, ditches, and composite monuments that were constructed to house the remains of the dead and provide a focus for community activities. Indeed, during the period in question, these monuments 'filled in' upland landscapes, changing communities' experiences of these regions and acting as highly visible and powerful signifiers. Directly preceding the period of large-scale 'domestic' and land boundary construction in many regions of Britain, these sites objectified their builders' changing relationship to the landscape at a crucial point in British prehistory (Owoc 2000, 2001a).

Names are powerful things, and the now-familiar terms used above to describe these monuments have, over the years, served in many ways to pre-define them, and condition the archaeological experience of them. Given this situation, it is somewhat ironic that the majority of interpretations of these sites and their significance have bypassed the material and constructional elements implied by these names, in order to concentrate primarily upon the cremated or inhumed individuals who occupied central or peripheral positions in relation to the monuments, or to each other. The significance of these sites, then, as re-made earth and stone, has been somewhat eclipsed by observations that the later Neolithic and earlier Bronze Age mark the appearance (for the first time) of true 'burial' on a large scale in British prehistory (Barrett 1990, 1994; Garwood 1991; Woodward 2000). The importance of this observation, and the studies that have been inspired by it, have lent much to our understanding of the period in question. This particular contribution is offered here not to critique this line of research, but instead to highlight the physical nature of the sites themselves and their role in the reproduction of Bronze Age society.

The archaeological narrative of the British late Neolithic and earlier Bronze Age has been derived (like those of other chronological periods) largely from analyses of central southern or northern sites. The archaeological record of the south-west during this period makes interpretative constructions based on Bronze Age burials less straightforward to produce, for several reasons. First, on the most practical level, the acidity of many south-western upland soils has long removed the traces of many inhumations (though a number of cremations do remain). The following record from the 1756 fieldnotes of William Borlase during excavations on St Mary's, Isles of Scilly (quoted in Marsden 1999: 118) parallels the experiences (if not the reactions) of many south-western barrow investigators past and present who, expecting bones and artefacts, found only stone and soil: 'In the first [entrance grave] we found no bones, or urns, but some strong unctuous earth which smelt cadaverous … there was a passage into it … it was walled on each side with masonry and mortar.'

Secondly, one's initial encounter with these sites – as embodied viewer in the landscape or excavation report reader – immediately indicates that in terms of their size, particular structural histories, and related ritual activities, many far exceed the practical requirements for any burials they overlie, contain, or are associated with. Thirdly, a long history of examination from the antiquarian period onwards has indicated that burial (in either a strict or general sense) was not always a primary or central activity at most of these sites. What is in fact far more compelling about the many late Neolithic/earlier Bronze Age 'funerary' sites of the south-west (and indeed elsewhere) is not their human remains at all, but the creation and composition of the sites themselves.

Although quite variable, most of the sites display one or more of a group of structural elements commonly used by builders of this period in the south-west and elsewhere, which include a mound or cairn (consisting of stone, earth, or some combination of the two), some type of ditch, and an earthen, stone or composite ring (**Figure 6.2**).

The mounds/cairns are often stratified, displaying multiple phases of construction that are often separated by a hiatus, artefact deposition, or other activities. Constructional activity at the sites was sometimes preceded by a number of other practices involving (among other activities) the erection of timber circles, the digging of pits, artefact deposition, the placement of human remains, and participant movement. When the sites are considered singly, the sum total of these activities presents various puzzling narratives which, when considered collectively, appear to overlap in only the most general of ways (eg sites generally appear as 'mound' or 'cairn' structures, surrounded by 'ring ditches', and contain 'inhumation' or 'cremation' burials, etc). It is suggested here that focusing upon the use of soil/stone as material culture in practice-centred, localised studies enables this motley plethora of activities to be resolved into truly meaningful patterns of Bronze Age funerary/ritual practice.

Figure 6.2: The site of Lower Innisidgen, St Mary's, Isles of Scilly, consists of an earthen mound supported by an outer granite wall and an entrance/chamber capped with large granite slabs.

REMAKING THE LAND

The mineral and geomorphological context for these varied constructions deserves a brief mention, since the structural and mineral diversity of the sites, and their local and regional character, directly relate in most cases to the varied geological composition of south-west Britain. The geological stratigraphy of the south-west is characterised primarily by Devonian Old Red Sandstone and various Carboniferous deposits that are cut by later, largely granitic intrusions (Selwood *et al* 1998; Rayner 1981). Other intrusive igneous rocks occur around the margins of the granite, and on the Lizard Peninsula in southern Cornwall. Towards the east, the older rocks disappear under a Permian and Triassic Marl and Sandstone cover. A Cretaceous Upper Greensand deposit also occurs in south-east Devon. As a result, depending upon their location, the sites in the south-western region incorporate a variety of materials, including a range of variously coloured conglomerates, gabbros, granites, flints, marls, quartzes, quartzites, sandstones, slates, siltstones, serpentines, and limestones. Weathering and glacial activity have given rise to a wide variety of variously coloured and textured superficial deposits overlying these rocks. Sub-surface sediments are either invisible, or visible as traces in ploughed fields, or dramatic exposures along coasts or in stream valleys. In exposed upland or coastal regions, the bedrock is encountered as protruding weathered cliffs, cairns, or tors, isolated boulders, or smaller scattered surface or beach materials. In such localities, one has the powerful experience of being dwarfed by stone or earth. Most mound, cairn or ring components were derived from local rocks and their overlying sediments and soils (through soil scrapes or turf cutting, ditch excavation, or surface material collection). In some cases, however, specially selected stripped, gathered or quarried materials were brought from a distance. While the particular arrangement and use of these various materials on the sites can be understood to be a function of local building traditions, the rationale informing these activities cannot be entirely explained or understood by this observation and thus necessitates deeper inquiry (Owoc 2000, forthcoming; see examples in Tilley 1999: 185–238).

As indicated above, expediency and practicality are not appropriate words to describe the construction practices of these south-western builders. Indeed, the 'strangeness' (to contemporary Western observers) of some of the creations suggests a very particular, localised, and in some cases temporally variable view about the appropriate use of particular materials. On many sites, mound or cairn construction was preceded by a careful turf and subsoil strip, or ditch creation. Such actions were clearly designed to reveal brighter subsoil components underlying darker surface elements. For example, many of the sites at the Davidstow Moor Cemetery near Bodmin Moor (Christie 1988) were constructed so that the underlying yellow subsoil stood out against the overlying, darker turf as rings, and in ditches and stripped circles (elaborated below). Further, mound and cairn construction often involved a timed and ordered employment of distinctive on- or off-site stratigraphic components in ways that indicate that distinctions were being made between various materials on the basis of their colour, texture, or vertical location in a natural stratigraphic arrangement. Notable examples of such practices include many of the St Austell Granite sites in Cornwall (Miles 1975), Carvinak in Cornwall (Dudley 1964), the East Putford sites in Devon (Ralegh Radford and Rogers 1947) and the Charmey Down/ Lansdown sites in Somerset (Grimes 1960; Williams 1950). The extent to which such practices involved careful soil choice, knowledge, planning and timing is revealed by the construction history of Upton Pyne 284b (located just north of Exeter, Devon; Pollard and Russell 1969).

Before constructing this multi-component mound incorporating some urned and unurned cremations, mourners at Upton Pyne stripped the site of its upper components and retained each separately (turf/topsoil, a thin leached lower A-horizon, and some orange/pink sandy subsoil). Over the course of some time, they then constructed a layered mound by placing these materials in what appears to have been a deliberate order – contrasting the orange reds with the darker turf and the whiter A-horizon, which itself was left to harden (**Figure 6.3**).

At a number of other sites exhibiting vertical differentiation of components in mounds, banks and revetments, the normal stratigraphic positions of the soils/sediments were often reversed, creating in effect either an upside down image of the buried landscape, or a contrasting paired arrangement. At the site of Trelen 2 (on the Lizard Peninsula in west Cornwall; Smith 1984), for example, mourners first created a ring of yellow subsoil, which they placed directly over

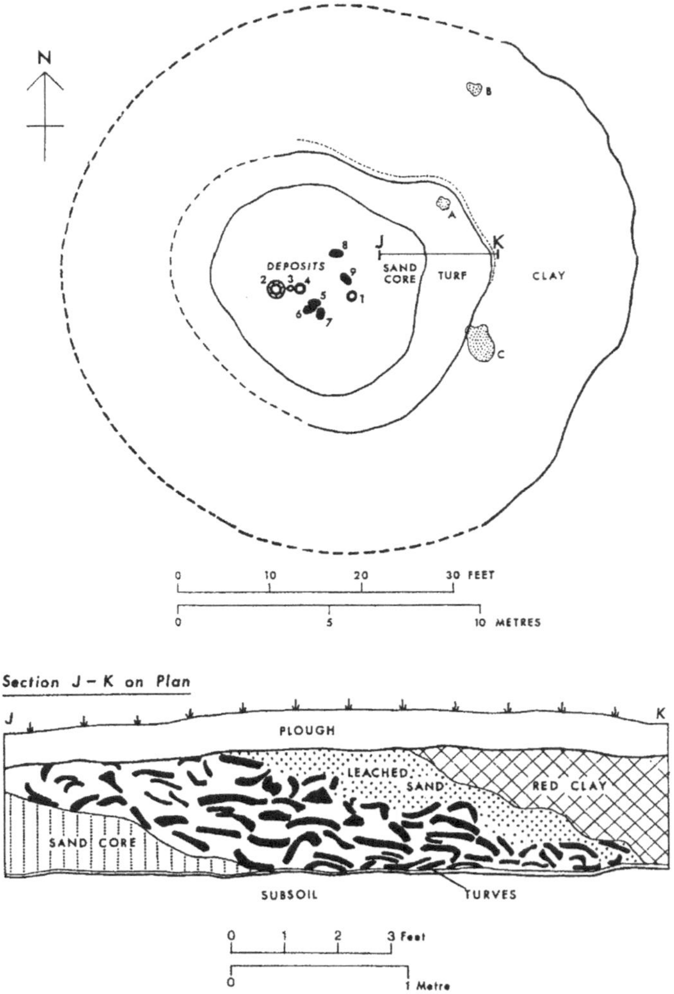

Figure 6.3: Stratified mound construction at Upton Pyne 284b (from Pollard and Russell 1969: 53, 55).

the natural ground surface. They then buried this ring with a second ring composed of turf and topsoil, in effect first reversing, then restoring the natural stratigraphy visible in the surrounding ditch from which they obtained both materials. By so doing, they brought what was invisible (below) to visibility (above), and then reversed that process. Additionally, over the course of a particular ritual sequence accompanied by site modification, many sites took on a lighter appearance or firmer composition as various clay, shale or quartz caps covered earlier sand or loamy mounds. In some instances, this sequence was initiated by the erection of *wooden* elements surrounding or encircling human remains, effecting a physical site transition from the impermanent organic to the permanent inorganic as subsoils and/or stones replaced rotting or removed timber uprights or deposits. Such a transition may have been a visible reminder of the changing physical nature and essence of the deceased and community throughout the funerary rite of passage. The evolving form of the site would parallel the shift from pliable fleshy person to hard mineral-like bone ancestor, and restore the permanence of the living community united with their ancestors after a social rupture caused by death (Owoc 2000).

A particularly noteworthy Cornish site, Crig-a-Mennis (Christie 1960), illustrates the changing nature of sequential mound components. The site began as a small turf mound that was subsequently capped with a loose pink sandy loam. Following this, two inurned cremations were placed next to the mound. Then, during incorporative post-burial activities, the mound was further augmented by the following additions: a firmer orange/gray loam, a local pinkish brown subsoil, shale rubble, a quartz and slate revetment wall and, finally, a white quartz cap. Interestingly, the builders of Crig-a-Mennis chose to locate the site directly over a geological break between soft pink sandstone and a harder greenish slate. On a practical level, this enabled them to take advantage of the two different loams overlying these materials. On a deeper level, given the fact that vertical and circumference boundary themes are consistently reinforced in these passage sites, the choice of such a location may well have been related to the meaningful 'construction' of this *geological* boundary by the local community (for a more thorough discussion of boundary themes in these sites see Owoc 2000, 2001b and below).

A concern with certain stratigraphic relationships is also evidenced by the excavation and refilling of particular pits located in, around or under mounds or enclosures. For example, at several Beaker sites in the Broad Down/Farway complex in south-east Devon, certain pits were filled with previously stripped topsoil loam, an underlying leached clay-with-flint subsoil, a local yellow clay subsoil, and a foreign, stoneless gray clay. The arrangement of these materials with respect to each other was generally consistent, and appeared to be governed by two distinct rules of stratigraphy that both highlighted the gray and leached clays and formed a limited arrangement of pit fills overall (Pollard 1967). If two or more of the materials were combined, the clay-with-flints always lay above the loam and foreign gray clay, and often formed a cap to the pits. The gray clay always occupied the lowest position when combined with any other material. This manipulation, which contrasted local and foreign materials and altered the natural stratigraphic order of things, finds many parallels in the built monuments of this group and elsewhere.

Finally, it is worth mentioning at this point that the deliberate selection and manipulation of particular soils or stones in mound/cairn/ring creation was part of a far wider 'construction' of the natural world that included regional viewscapes and the cosmos. For example, the depositional timing or location of human remains, mound or enclosure materials often occurred with reference to important solstial and temporal transitions such as the midsummer and midwinter sunrises (MWSR) and sunsets, and/or cardinal directions (Owoc 2000, 2001b). In such cases, the builders appear to have set up arenas where meaningful links between the living and the dead, the buried landscape, the viewscape, and the cosmos could be established.

From these examples and many others, one is left with the overwhelming impression that the creators of the south-western sites were highlighting and deliberately remaking the natural surface (and buried) landscape in very particular sorts of ways. Further, the principles governing these activities appeared to rely heavily upon their creators' embodied and sensual encounters with the vertical and horizontal landscape/viewscape, and the colours, luminosity, textures, and hardness/softness of the mineral world.

MEANING AND THE MINERAL WORLD

Since direct or consistent relationships cannot be demonstrated between the form and size of these monuments and the social identities and status of their builders or the personages within them, questions regarding the form and meaning of these carefully crafted and structurally elaborate creations remain. Recent work on the signifying character of the prehistoric landscape throughout southern Britain (eg Bradley 2000; Parker Pearson and Ramilisonina 1998; Tilley 1994, 1995, 1999), and ethnographic accounts of indigenous mineral taxonomies (eg Arnold 1971; Ollier *et al* 1971; see also Boivin, Brumm, Taçon, and Saunders, this volume) indicate the likelihood that components of the south-western buried and surface landscape were part of a broad symbolic system involving many materials and places. Following the work of Barth (1975, 1987), Bourdieu (1977) and Turner (1967, 1969) on the creation of symbolic taxonomies, it has been suggested that the temporal or spatial juxtaposing of monument materials with other important dominant symbols on the sites may have served to create one or more pairs of meaningful analogous oppositions (Owoc 2000, forthcoming). Such oppositions, created in action and objectified materially, could have formed the basis of partly linked, partly autonomous practical taxonomies informing both the daily and ritual practices of south-western Bronze Age communities. Manipulated at times of important transitions in the life of the cosmos and the community, soil and stone would simultaneously have been active in the creation and reproduction of their own signifying qualities, and instrumental in the conceptualisation, acceptance and facilitation of these transitions.

The site of Davidstow Moor Site III, along with some of its neighbours, will serve to illustrate how such a process of meaning creation may have taken place (**Figure 6.4**).

This site lay within a slowly developing, late Neolithic through early Bronze Age cemetery and ritual landscape on Davidstow Moor, in the northern part of Bodmin Moor, Cornwall. The cemetery, excavated in its entirety during the 1940s, was characterised by a number of distinctive yellow subsoil ring banks and turf strips, among which also lay yellow pond barrows and yellow revetted turf mounds (Christie 1988). Throughout the period in question, judging from the structure of the sites themselves and the arrangement of their interior and exterior features, the local community appeared to have been highly interested

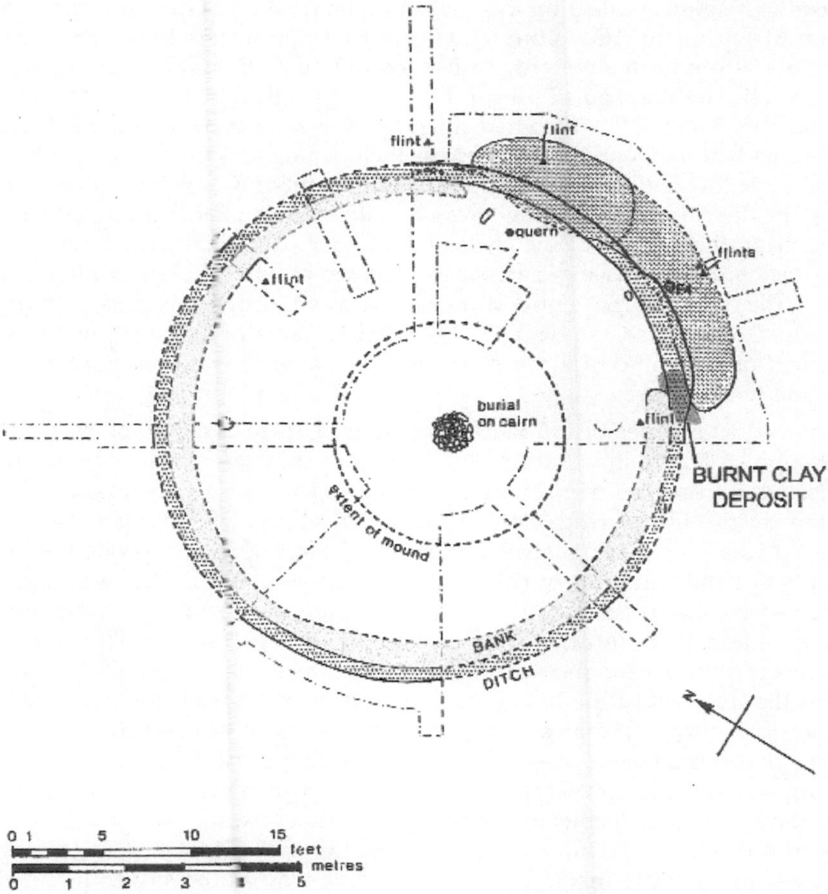

Figure 6.4: Davidstow Moor III (after Christie 1988: 58). Lighter shading represents yellow and white clays, while darker shading indicates burnt clay at entrance.

in circular solstial movement and seasonal timing. This was evidenced by a common concern to mark certain midsummer or midwinter sunrises and sunsets with various features, posts, stones, and enclosure entrances. Local soils were also employed, as yellow subsoil circles, rings or mound revetments mimicked the sun, or its passage around, above and below the horizon via metaphors based on analogous colours and spatial properties (Owoc 2000, and below). While no two sites were alike, each echoed one or more of its older or younger neighbours in the arrangement of its features, or the form of its mound, ditch or circle. Site III may well be the result of a very long-term series of activities that began with the construction of a late Neolithic causewayed bank enclosure. Later, funerary activities and structural modifications to the site took place. In the initial phase of construction, the builders first divided the area of the site into two zones though the creation of a yellow subsoil ring with a turf core (materials which they obtained from a ditch just outside the ring). The excavation of the two contrasting materials from the ditch, and the creation of the ring would have drawn attention to the colour and brightness contrasts between the two materials while both reversing and highlighted their natural stratigraphic relationship. The ring had a 5 m gap to the east/south-east. In the later phase of the site, after some of the ditch had silted up, this gap was subsequently blocked with a small mound, and the ditch across the opening was filled with a colourful spread of yellow and white clay subsoil. Some of the clay was heated or burned, giving the deposit a mixed red, white and yellow colour. For entrance and egress to and from the interior of the enclosure, builders left a narrow passage at the south-east end that was aligned near the horizon point of the midwinter sunrise when observed from the centre of the site. It was through this passage that the mourners transported the bagged, cremated remains of an adult, whom they deposited on a small central cairn. A small turf mound was then built over the cairn and the deceased.

Viewed as a series of intimate encounters with the viewscape and buried landscape of Davidstow Moor, the unusual practices at Site III transform themselves into a series of meaningful strategies. Through a principle of *enclosure* created by the initial ring/ditch and the subsequent blocking mound, the deceased was placed into a spatial and conceptual category opposite the living members of the community (living:dead::inside:outside). This was actively reinforced by the transfer of the remains through the entrance passage by the mourners and their return to the outside of the enclosure. Further, by the addition of the turf mound, the deceased was also placed below the surface (under the turf) whilst the living remained above it (above:below::living:dead). The yellow clay-turf bank, with its contrasting coloured and reversed *stratigraphic* arrangement, may have elaborated this above:below opposition by linking it to the physical qualities of the materials themselves (light:dark::visible:invisible::above:below). Further, the use of the yellow ring bank and ditch by the mourners at Davidstow Site III served to frame the activities taking place inside in much the same way as the sun in its passage around the horizon framed the world of this community. Through the skilful manipulation of soil as a material culture symbol, the builders appear to have created an *imago mundi* (Eliade 1959: 42–43.58) – a miniature universe that served to link the substance of the funerary ritual with the larger functioning and structure of the cosmos itself. The colourful passage, with its *axial* orientation on an important daily and seasonal transition (a sunrise and yearly solar shift) may have acted as a crucial condensing symbol linking all the classification schemes set up at the site. Through this passage, the living and the deceased could be seen as entering or emerging from the world of the dead. Further, its orientation connected the enclosure scheme or plane to the axial one, since the entrance lies near the two-fold cosmic transition between darkness and light (sunrise) and the December ending/beginning of the solar cycle. Locating the entrance in this area enabled the passage between the spaces and realms of the living and the dead to be linked to: (1) the vertical passage of the sun between two cosmic planes, and (2) the temporal solar transition in midwinter (which is both an ending and a beginning). The colourful, bright appearance of the burnt subsoil in the entrance passage highlighted the importance of this location, but also, like the ring bank, may have served as a reminder of the symbolic qualities of the subsoil. This material shares colour properties with the sun, both before and after thermal alteration, and on the site both naturally (in the ditch) and artificially (through the entrance deposit and the ring bank) mimics the sun's changing position below and above the horizon. From a simple series of carefully structured oppositions, then, a complex scheme of perception could have been created by the activities at Site III through the meaningful construction of space, time and the mineral world:

Enclosure	***Axial***	***Stratigraphy***
inside/outside	(MWSR+coloured clay)	turf/yellow clay
dead/living	above/ below	above/below
	light/dark	dark/light~light/dark
		living/dead

Considered within the context of a life crisis ceremony at the site, the establishment of links between several cosmological states of being and space/time may have allowed this community of mourners to (1) situate and alleviate the social rupture of death within a wider understanding of the natural order of things, and (2) objectify and reproduce important community traditions of knowledge through action and material culture.

THE COMMUNITY IN A MINERAL WORLD

The title of this chapter, and the above account, appear to rest on a philosophical conception of the relationship between humans and landscape in which an embodied actor in the material world at once constructs that world and is in turn constructed by it. This description, however, is not entirely sufficient to describe that experience, or the creation and reproduction of Bronze Age traditions of knowledge in south-west Britain. Archaeological accounts of the meaningful relationships between people and the landscape often operationalise a notion of phenomenology that privileges the *individuality* of experience (McGuire and Tuchanska 2000). This is problematic since such a notion rests on an ontological structure of being that arises primarily from a single human-world or human-material relation. Alternative conceptualisations of being, knowing and acting articulated in practice theory, and recent reflections on the work of Heidegger suggest that 'being' is always a 'being-together', or a 'being-with-others-in-culture'. Further, it is argued that understanding about self and world arises not merely through human relations to things but through *participation in communities* (McGuire and Tuchanska 2000: 72, 103–10; Schutz 1969). Human dealings with the physical world, therefore, are better formulated and studied as mutually engaged practical activities, or as elements of collective, shared practices (Barnes 2000: 64–66; McGuire and Tuchanska 2000: 87, 100). Such conceptualisations not only require recognition that communities have an existence beyond that of the sum of individuals – they also necessitate an inquiry into their constitution and reproduction (McGuire and Tuchanska 2001: 110–11).

This context provides a far better way to conceive and write a phenomenology of the buried landscape for south-western Britain in later prehistory. The sites described above were quintessentially *community* monuments. The archaeological record of the late Neolithic and earlier Bronze Age south-west indicates overwhelmingly that these sites were the products of communal labour during protracted rites of passage and/or seasonal ceremonies. Further, they were visited and revisited by local communities of mourners or celebrants who altered their physical forms through (amongst other activities) structured deposits of carefully selected earth and stone. These mourners simultaneously experienced those forms at a number of physical and perceptual levels and, moreover, participated in a negotiated *social* construction of their meanings. The Bronze Age entrance grave monuments on Isles of Scilly granite promontories, for example, were observed and repeatedly visited over many years by local families who, during constructional alterations and the periodic, restricted deposition of cremated remains and ceramic vessels, observed a panorama of natural, though similar granite outcrops, as well as the granite entrance graves and sepulchral activities of their regional neighbours (**Figure 6.5 colour section**; **Figure 6.6**).

Practices like site construction, alteration and movement relied upon 'shared members' knowledge' (Barnes 2001: 21). This knowledge informed participants not only about social identity, power and the particular relationships in which these things were established and perpetuated, but also addressed the mythical world where all ontological and existential questions were explained. These explanations relied on a common 'trans-individual resource' (Bauman and May 2001: 29) of symbolic and metaphorical knowledge about the world and its past that was objectified and reproduced through the active engagement of these people with the mineral material world, their monuments, and the remainder of their surrounding environment.

Figure 6.6: View of late Neolithic/early Bronze Age entrance grave monument (capstone and chamber) on Tregarthan Hill, Tresco, Isles of Scilly, UK.

CONCLUSION

It has been suggested that the definition of material culture would be well augmented by an appreciation of the myriad ways in which humans actively and meaningfully use soils, sediments and stones in their daily encounters with the world. An expanded definition shifts our conception of minerals from 'raw' materials forming the background for human activities, to constructed, signifying elements within individual and community epistemological projects. The monument data from south-west Britain indicate that questions about how people in the past interacted with the mineral world must also extend below and beyond the ground surface. Finally, the nature of these monuments as sites of community action and reproduction lends support to the assertion that our archaeological appreciation of the human-landscape encounter would be benefited by revising current phenomenological constructions in archaeology.

Acknowledgments

I wish to thank the Devon Archaeological Society and the Cornish Archaeological Society for permission to reproduce Figure 6.3 and a slightly modified version of Figure 6.4, respectively. David Pedler deserves special thanks for editorial assistance during the completion of this paper, and for the preparation of the figures. Jacky Nowakowski offered many insights concerning Cornish barrows and their excavation during the preparation of this paper, and Mike Parker Pearson inspired my plunge into mineral meanings many years ago. Thanks also go to Nicole Boivin for her editorial advice.

References

Arnold, DE (1971) 'Ethnomineralogy of Ticul, Yucatan potters: etics and emics' 36(1) *American Antiquity* 20–40

Barnes, B (2000) *Understanding Agency: Social Theory and Responsible Action*, London: Sage

Barnes, B (2001) 'Practice as collective action', in Schatzi, TR, Cetina, KK, and Von Savigny, E (eds), *The Practice Turn in Contemporary Theory*, New York: Routledge

Barrett, J (1990) 'The monumentality of death: the character of Early Bronze Age mortuary mounds in Southern Britain' 22(2) *World Archaeology* 179–89

Barrett, J (1994) *Fragments of Antiquity*, Oxford: Blackwell

Barth, F (1975) *Ritual and Knowledge among the Baktaman of New Guinea*, New Haven: Yale University Press

Barth, F (1987) *Cosmologies in the Making*, Cambridge: CUP

Bauman, Z and May, T (2001) *Thinking Sociologically*, Oxford: Blackwell

Bourdieu, P (1977) *Outline of a Theory of Practice*, R Nice (trans), Cambridge: CUP

Bradley, R (2000) *An Archaeology of Natural Places*, London: Routledge

Christie, P (1960) 'Crig-a-Mennis: a Bronze Age barrow at Liskey, Perranzabuloe, Cornwall' 26 *Proceedings of the Prehistoric Society* 76–97

Christie, P (1988) 'A barrow cemetery on Davidstow Moor, Cornwall' 27 *Cornish Archaeology* 27–169

Deetz, J (1977) *In Small Things Forgotten: The Archaeology of Early American Life*, New York: Anchor

Dudley, D (1964) 'The excavation of the Carvinak barrow, Tregavethan, near Truro, Cornwall' 2(44) *Journal of the Royal Institute of Cornwall* 414–51

Eliade, M (1959) *The Sacred and Profane: The Nature of Religion*, Trask, WR (trans), New York: Harper and Brothers

Garwood, P (1991) 'Ritual traditions and the reconstruction of society', in Garwood, P, Jennings, D, Skeats, R and Toms, J (eds), *Sacred and Profane*, Oxford: Oxford University Archaeology Monograph 32

Grimes, WF (1960) *Excavations on Defense Sites 1939–1954*, London: HMSO

Herskovitz, M (1963) *Cultural Anthropology*, New York: Knopf

McGuire, JE and Tuchanska, B (2000) *Science Unfettered: A Philosophical Study in Socio-historical Ontology*, Athens: Ohio University Press

Marsden, BM (1999) *The Early Barrow Diggers*, Stroud: Tempus

Miles, H (1975) 'Barrows on the St Austell Granite' 14 *Cornish Archaeology* 5–81

Ollier CD, Drover, DP and Godlier, M (1971) 'Soil knowledge amongst the Baruya of Womenara, New Guinea' 42(1) *Oceania* 33–41

Owoc, MA (2000) *Aspects of Ceremonial Burial in the Bronze Age of Southwest Britain*, unpublished PhD thesis, University of Sheffield

Owoc, MA (2001a) 'The times, they are a changin': experiencing continuity and development in the Early Bronze Age funerary rituals of southwestern Britain', in Brück, J (ed), *Bronze Age Landscapes: Tradition and Transformation*, Oxford: Oxbow

Owoc, MA (2001b) 'Bronze Age cosmologies: the construction of time and space in south-western funerary/ritual monuments', in Smith, AT and Brooks, A (eds), *Holy Ground: Theoretical Issues Relating to the Landscape and Material Culture of Ritual Space*, BAR International Series 956

Owoc, MA (forthcoming) 'Beyond geoarchaeology: pragmatist explorations of an alternative British Bronze Age viewscape', in Robertson, J, Siebert, J, Fernandes, D and Zender, M (eds), *Space and Spatial Analysis in Archaeology*, Calgary: University of Calgary Press

Parker Pearson, M and Ramilisonina (1998) 'Stonehenge for the ancestors: the stones pass on the message' 71 *Antiquity* 308–32

Pollard, SHM (1967) 'Seven prehistoric sites near Honiton, Devon. Part I' 25 *Proceedings of the Devon Archaeological Society* 19–39

Pollard, SHM and Russell, PMG (1969) 'Excavation of round barrow 248b, Upton Pyne, Exeter' 27 *Proceedings of the Devon Archaeological Society* 49-78

Ralegh Radford, CA and Rogers, EH (1947) 'The excavation of two barrows at East Putford' 3 *Proceedings of the Prehistoric Society* 156–63

Rayner, DH (1981) *The Stratigraphy of the British Isles*, Cambridge: CUP

Schlereth, TJ (ed) (1969) *Material Culture Studies in America*, Nashville: American Association of State and Local History Press

Schutz, A (1969) *The Phenomenology of the Social World*, Walsch, G and Lehnert, F (eds), Evanston: Northwestern University Press

Selwood, EB, Durrance EM and Bristow, CM (1998) *The Geology of Cornwall and the Isles of Scilly*, Exeter: Exeter Press

Smith, G (1984) 'Excavations on Goonhilly Downs, the Lizard, 1981' 32 *Cornish Archaeology* 3–42

Tilley, C (1994) *The Phenomenology of Landscape*, Oxford: Berg

Tilley, C (1995) 'Rocks as resources: landscapes and power' 34 *Cornish Archaeology* 5–57

Tilley, C (1999) *Metaphor and Material Culture*, Oxford: Blackwell

Turner V (1969) *The Ritual Process: Structure and Anti-Structure*, Ithaca: Cornell University Press

Turner, V (1967) *The Forest of Symbols*, Ithaca: Cornell University Press

Williams, A (1950) 'Bronze Age barrows on Charmy Down, Landsdown, Somerset' 30 *Antiquaries Journal* 34–46

Woodward, A (2000) *British Barrows: A Matter of Life and Death*, Stroud: Tempus

BEYOND TEXT

As the chapters in this volume make clear, the symbolic power of the mineral world derives in large part from the sensual, evocative properties with which soils, stones and other mineral elements are endowed. While many of the chapters' authors attempt to communicate something of this evocative quality, it may be argued that, however powerful, words nonetheless fail to capture the full sensual and emotional impact of mineral entities. For this reason, some of the more visually evocative figures that accompany individual chapters in this volume have been gathered here within a separate section of colour figures. The purpose of this section goes beyond that of simply illustrating in colour form academic points described by individual authors. The section has instead been devised primarily with the aim of communicating, through an additional, more sensual and less cerebral medium, something more of the aesthetic, evocative, expressive and material power of the mineral world. While still engaging only one dimension of a complex sensual experience, it nonetheless plays an important role in communicating one of the primary messages of the volume: that the mineral and indeed entire material world is powerful not only because it represents mental concepts, but also, perhaps more importantly, because it addresses at one time a whole variety of human modes of experience, sensation and imagination. The images that follow capture something of the visual dimension of this unique capacity.

re 2.1: Massive quarry 'The Ochre Pits', west of Alice Springs, Northern Territory.

Figure 2.3: Eastern Kunwinjku people of central Arnhem Land painted with white pipe clay after gathering pigment from nearby rock-art sites.

Figure 2.5: Panel of polychrome rock-paintings, Kakadu National Park.

re 5.1: View from inside Beehive Shelter (© Rick Bury).

re 5.2: Remnant smears of red pigment at Rattlesnake Shelter (© Rick Bury).

Figure 5.4: Fingerprints from the direct application of red pigment at Honda Ridge (© Rick Bury).

Figure 5.5: The reflective surface at the Honda Ridge rock art site (© Rick Bury).

Figure 6.5: View of late Neolithic/early Bronze Ag entrance-grave monument (capstone visible in foreground) and granite outcrop (above) on South Hill, Samson, Isles of Scilly, UK.

Figure 7.1: Aztec wooden mask with turquoise inlay. © Reproduced with permission of the Werner Forman Archive.

re 8.3: Rock art depiction of Namarrgon the Lightning Man, painted by Najombolmi or 'Barramundi' lie in the 1960s, Kakadu National Park (photo courtesy Paul Taçon).

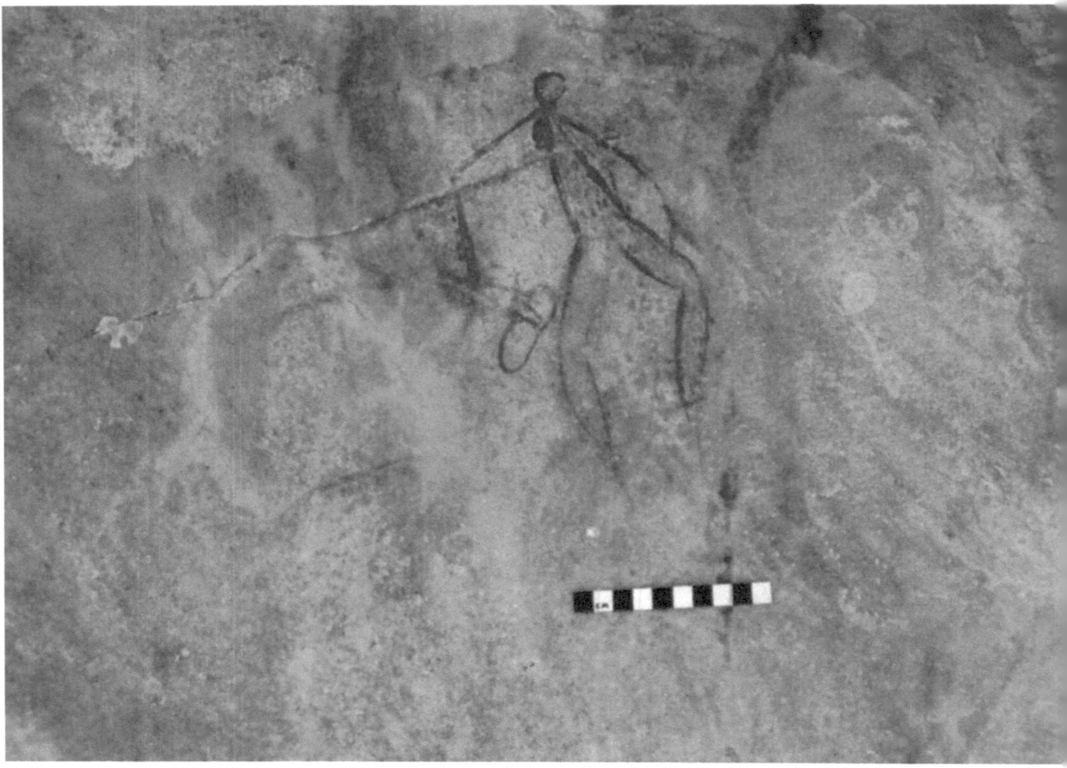

Figure 8.4: Ancient Dynamic tradition painting of a female figure holding a large stone axe, western Arnhem Land (photo courtesy Paul Taçon).

Figure 9.3: A village boy displays samples of the soil types that have been used to plaster and decorate the fl below. *Pili mitti* is the red soil on the right, and *safed mitti* is the white soil on the left.

re 9.6: Cooking area of a Balathal house, showing *kotha* (storage bin) on the left and hearth in the right er. The floor and the area above the hearth are plastered with red *pili mitti*.

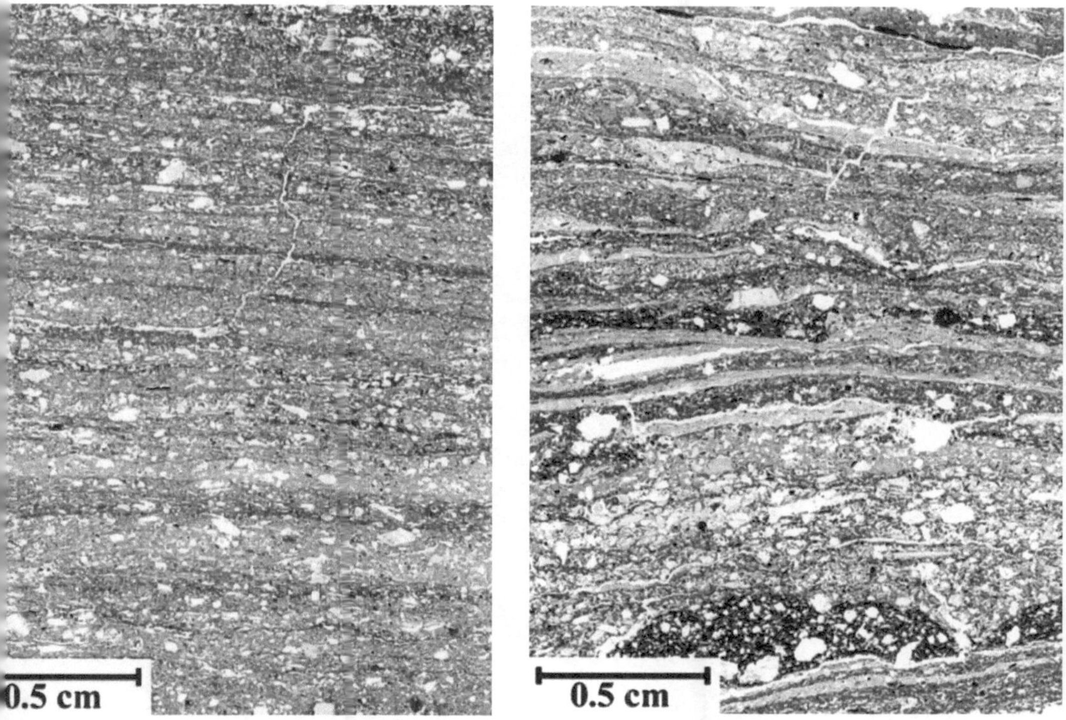

re 9.7: Micromorphological comparison of plasters used in different parts of the house. Plaster layers from loor of a regular, non-sacred area (a) are contrasted with those from the surface of a *puja* mound (b), which ides many layers of brighter red and white plaster.

Figure 10.3: The northern passage grave of Les Mousseaux, looking outward along the southern side of the entrance passage. Note the quartz block among the dry stone walling supporting the capstone.

Figure 11.1: *Bibemus Quarry,* Paul Cézanne, BF #34 © reproduced with the permission of The Barnes Foundation™, all rights reserved.

– 7 –

THE COSMIC EARTH

MATERIALITY AND MINERALOGY IN THE AMERICAS

Nicholas J Saunders

Throughout the Americas, between at least 3000 BC and AD 1492, cultural values and spiritual qualities inhered in the diversity of matter that constituted the natural world. For pre-Columbian and Native American societies, identity and energy flowed between what Western taxonomies regard as different levels of physical and metaphysical existence – sky, land and sea, and life and death – binding the power of ancestors and immanent cosmic forces to the everyday lives and bodies of human beings through objects found and altered on earth.

In the holistic worldview of indigenous Amerindian peoples, what Western science considers the 'natural landscape' was regarded as a cultural landscape bound together, shaped and given meaning by the presence of spiritual power and mythology. The geological components of cultural landscapes were not inanimate physical matter, but rather were imbued with cosmological significance. Toponomy recapitulated cosmogony – physical places featured in creation myths and represented the past acts of supernatural beings in the present. This cosmological and spiritual significance found social expression in the physical possession and ritual wielding of such varied geological matter as quartz crystals, flint, greenstones, turquoise, ochre and obsidian, and in their multi-media somatic use as symbols of social status and spiritual and ethnic identity.

The pan-Amerindian conception of the earth as a living entity, and of its various geological elements as possessing specific spiritual, ideological and technological qualities, invites an anthropologically-grounded multidisciplinary investigative approach. The ways in which cultures conceive and articulate the relationships between their ideas of a 'sacred earth' and its individual geological components (and their use) can be studied in many ways. My aim here is to explore but one aspect of these relationships: how different Amerindian cultures have conceived aspects of the physical world which we call geological, and how these aspects have been symbolically acquired and utilised in the social arena.

The different technological properties, colours, textures and sensual qualities of different kinds of geological matter were perceived according to local and regional classifications, but often their sheen and lustre suggest a shared status as brilliant embodiments of the cosmically charged fertilising power of light (Saunders 1998, 1999). Sacred substances transformed into cultural artefacts were embodiments of religious beliefs, and the technological process chosen to work the material was itself charged with spiritual and ideological force. Technological choices drew metaphysical and metaphorical connections between those who made, used and wore the finished items and the multi-sensorial nature of the 'raw' material (see Houston and Taube 2000).

The working of minerals was more than craftsmanship – it represented the manipulation and recombination of dangerously ambivalent spiritual essences by processes that trapped and concentrated cosmic energies in cultural forms. The origins of metalworking – an undoubted technological advance but so often misrepresented as a break with prior cultural practices and beliefs – cannot be understood as standing alone, or outside the realms of indigenous natural philosophy. Rather, metalworking was an extension of millennia-old beliefs and practices that saw form and essence transformed by fire, technology and ritual knowledge (Helms 1986; Hosler 1994; Lechtman 1993; Saunders 2003: 23–26).

In this chapter, I offer a brief overview of the role of geology and geography in a worldview that was, and in part remains, a fusion of what in Western understanding are indissolubly separate spheres of natural phenomena and cultural activity. What emerges from the many different kinds of evidence discussed here is a view of a cultural world composed of complex and nuanced ideas concerning the nature and role of various kinds of minerals and landforms that challenge simplified Western conceptions of nature/culture distinctions.

As in several recent publications (Saunders 1999, 2001, 2003), I explore how aspects of the natural world were symbolically acquired for the construction of indigenous natural and moral philosophies (ie a worldview or *cosmovision*). In particular, I aim to show how various indigenous ways of seeing and knowing the world through its mineral nature linked the human body with society, its mythic constructions and the natural world that surrounded and inspired them. Through myth and analogical symbolic reasoning, connections reached out to the farthest limits of human imagination, so that, for example, the wearing of mineral jewellery could be seen as carrying the cosmic fertilising energies of the sun, or tapping the powers of the watery realm of death and rebirth.

Through sheen, colour and technological attributes, the textured surfaces of the earth work their way up into material culture. Their attractions may be obviously functional, unrealised potentials or 'affordances' (Gibson 1986: 127–43), or perhaps 'simply' aesthetic (see Gell 1998: 1–5). Yet, from the moment minerals are picked up by people, they enter the circulatory systems of culture, where use-value, symbolic propensities, spiritual attributes and social value are all brought to bear, 'changing' the raw material even before it becomes an artefact. The artefact that is then made from it (and whatever its functional purpose) has the potential to reify cultural attitudes and valuations, and to carry an appropriate symbolic charge.

MOTHER EARTH: MINERALS AND SACRED LANDSCAPES

Conceptualisation of the earth as a sacred, female, living entity is widespread across the Americas. It is part of a worldview that sees the physical world and its natural phenomena as animated by, and infused with, spirit force.

The most salient features of Mother Earth are mountains – visually impressive, the source of water, generators of local weather, and the symbolic home of gods and ancestors. In Andean South America especially, indigenous peoples have long incorporated telluric symbolism into their natural and moral philosophies, perhaps not surprisingly in this dramatic and tectonically active region. Traditional Andean ways of sensing and representing the landscape in mythological terms have seen geography, geology and climate play a central and interconnected role in cosmogony and religion, offering detailed accounts of the supernatural origins of physical surroundings, and explaining how and why mountains, rocks, lakes, rivers and shrines were located in particular places, and how they became full of supernatural power.

Central to this way of understanding the Andean world has been the belief in transformation – the ability of culture heroes to change into rocks and stones, or of these inanimate objects to adopt human form, give supernatural help to humans, and then revert to their original shape. This magical quality of stone is one explanation why mountains, rocky outcrops, carved statues and columns, as well as stone-built temples and pyramids, have held such significance for Andean peoples, in the past and up to today.

Much work has been done on documenting and assessing the polysemic nature of Andean landscapes, especially mountains (eg Reinhard 1988, 1991, 1992, 1996). Mountains are sacred places and sacred beings (*huacas*), and in the Inka empire they were linked one to another in a system of snow-capped peaks that tied the empire together ideologically as well as geographically. Overlooking the Inka capital of Cuzco are two prominent mountains, the 'brothers' Salcantay and Ausangate. They are regarded as the 'fathers' of all mountains, and are often invoked in rituals that involve prayer to other smaller mountains for good weather and crop fertility. Salcantay is commonly referred to as *Apu* or Lord, and local people appeal to its power in order to cure illness and disease (Reinhard 1991: 13–15). Yet, like all spirit-mountains, Salcantay can turn against humans, by sending biting winds, snowstorms or sleet to damage crops and kill livestock. Those who today pass near take care to make offerings and placate its anger.

Andean beliefs about the supernatural nature of mountains and stones had an important impact on the location and style of Inka architecture. The huge granite blocks that the Inkas used to construct their civic and royal buildings themselves came from quarries located on the flanks of 'living' mountains. Temples, fortresses and palaces were all built from these sacred raw materials in a style that integrated cut-stone with natural outcrops, and defied the earthshaking tremors of the region (eg Gasparini and Margolies 1980).

The symbolism of different kinds of stones in the landscape was an omnipresent feature of Inka and wider Andean beliefs. Hills, mountains, caves and rocky outcrops (unaltered and shaped), as well as unusual mineral formations, were all considered sacred and worshipped. Such was the deep-rooted nature of this practice that Spanish priests made lists of the more obvious (ie shaped) stone objects that they regarded as the focus of pagan idolatry. Plain or coloured stones could be unaltered 'found objects', or carved into domestic cult items in the shape of llamas or alpacas and passed down as heirlooms (Hyslop 1990: 106–07). Some stones were regarded as *pacarinas*, symbols of ethnic origin, and served as magical protectors of people (*ibid*).

Inka beliefs concerning minerals extended from the monumental to the miniature, and even the microscopic. At the mountain-top city of Machu Picchu, the so-called Principal Temple had a layer of fine white sand scattered on its earthen floor. This parallels similar discoveries at several ceremonial plazas at Cuzco. This sand appears to have been brought from the Pacific coast by order of the emperor Pachacuti and was symbolically associated with the Inka creator deity Viracocha, and the ocean as 'mother' and ultimate source of all water and fertility (Reinhard 1991: 45).

The presence of sea sand in the Andes suggests a symbolic association of the sea with the surrounding mountains – themselves controllers of water, and whose worship included the ritual offering of sea shells also brought from the Pacific (Murra 1975: 257). Here, sand was the focus of a network of symbolic connections between large and small bodies of water, mountains and religious activities aimed at petitioning for rain and the fertility of crops and animals.

The focus of Inka stone worship was their imperial city of Cuzco – itself the empire's most important *huaca* and composed of many smaller stone *huacas*. Of perhaps 300 sacred places in Cuzco, about a third were represented by rocks – and some 14 of these appear to have been identified as *pururaucas* – stones that, in official creation myths, had become warriors to help the Inkas fight their enemies, the Chanka, at the beginning of their rise to imperial power (Garcilaso de la Vega [1609] 1987: 280–81).

Illustrating the sacred anthropomorphised nature of mountains and minerals in Andean religious thought are the beliefs and activities that surrounded gold and silver mining during Inka times. Many Inka-period mines were gallery mines, excavated into the flanks of the living body of mountains, and were regarded as holy places. Rituals were held before mining began, to petition the mountain to yield up its mineral wealth.

Both mines and their enclosing mountains were worshipped by the Inkas, each mine having its own 'idol' or *huaca*. According to Cristóbal de Albornoz ([c 1580] quoted in Berthelot 1986: 82–83):

> They selected the most beautiful rocks of these metals ... They worship them, calling them *the mothers of those mines*. Before going to work in the mine ... they worship the rock and drink in its honor, calling it the *mama* of the mine being worked.

These *mamas* were the most beautiful stones associated with the ores; in the case of gold, they were conglomerates in which flecks of gold were mixed with sparkling quartz or silica, and in the case of mercury, they were pieces of blood-red cinnabar (Berthelot 1986: 83). Together with the sought-after metals, these strange accompanying mineral formations added to the supernatural aura of the mine and its materials (*ibid*). Linking the upper and lower worlds of Inka cosmology, metals and minerals were 'the site of a ritual encounter between mountain and mine' (*ibid*: 84).

Today, similar beliefs survive throughout the Andes, though not unaffected by 500 years of Western contact. In Bolivia, Mount Kaata is conceived as a human body, with different locations named according to human anatomical features. Worldview and spirituality are demonstrated by the area's local diviners who live in the village of Kaata, identified with the heart, from where 'they can pump blood and fat, principles of life and energy, to the rest of the mountain's metaphorical body' (Bastien 1985: 37):

> The mountain has a head where alpaca hair and bunchgrass grow. The highland herders of Apacheta [village] offer llama fetuses into the lakes, which are its eyes, and into a cave, which is its mouth, to feed the head. ... Kaata is the heart and guts, where potatoes and oca grow beneath the earth. ... [The inhabitants say] If we don't feed the mountain, it won't feed us. (*ibid*: xix)

In Mesoamerica, similarly, 16th-century ethnohistoric sources indicate that mountains and volcanic peaks were sacred places, and also associated with water and fertility. In other words, physical landscapes were natural prototypes for cultural paragons in mythic narratives. The watery and fertile Valley of Mexico inspired the Aztec conception of paradise – a moisture-laden place known as Tlalocan, home to the rain god Tlaloc. Here, human beings took the form of 'flowers, birds, and shimmering gems' (Burkhart 1992: 89). Living landscapes and social life were associated also in metaphor, as in the Aztec term for village or community, which was *altepetl* or 'mountain of water' (Broda 1987: 93).

The relationship of prominent landscape features and their geological (mineral) constituency is seen clearly at the early Formative Period site of Tlapacoya in the Valley of Mexico. Dating to around 1250 BC, artefacts from the site include polished greenstone celts and a shiny iron-ore mirror. Here, geology, location and phenomenology may have been closely linked some 3,000 years ago. For, while today Tlapacoya is an isolated volcanic hill, in pre-Columbian times it was an island in the centre of Lake Chalco. According to Niederberger (2000: 175), 'the volcanic island of pink andesite ... reflecting in the lake would have constituted an impressive sight and may have been conceived, with its grottoes and springs, as a sacred place'.

Whatever meanings particular mountains or other geological features had for different cultures, they remained part of sacred landscapes that provided the raw material for one of the fundamental elements of cultural life across the Americas. As with other mineral formations, it may be that clay was seen as a 'magical' substance long before it was transformed by technology into ceramic, and perhaps different kinds of clay were held to be spiritually significant in their own right. They were mixed with other substances that were believed to impart different kinds of power and protection. The choice of clays and temper in some instances may have depended on the shininess of their mineral inclusions (Arnold 1993: 113; Lunt 1988: 493) – thereby incorporating into the vessel or figurine whatever qualities these sparkling minerals were believed to possess.

Sometimes, these substances were incorporated literally into human bodies, perhaps for physical as well as spiritual sustenance. Tschudi (quoted in Karsten 1968: 242) recorded that Amerindians of highland Bolivia ate the white clay called *pasa* from which they made their pottery and, interestingly, their ceramic figurines of Christian saints. The Shirianá of the upper Orinoco also consumed white clay kneaded into balls (Métraux 1948: 863), and the Otomac of the Venezuelan Llanos are reported as eating up to three-quarters of a pound of clay daily after bathing (Kirchoff 1948: 443). The Timbú, who inhabited the islands of the Paraná river, ate earth fried in fish grease (Lothrop 1946: 189). These scattered references may be given a degree of cosmological coherence by an example from Mexico, where the inhabitants of San Miguel Tzinacapan say 'We eat from the Earth/because of this the Earth eats us' (T Knab, quoted in López Austin 1988: 313).

While eating soil and clay was unusual, though widespread, consuming salt was arguably the most common form of ingesting minerals in the Americas. Attitudes toward salt illustrate the complex symbolism of this substance and the ways in which it overlapped spheres of economics, sustenance and cosmology. Together with emeralds, salt was the main trade mineral of the Colombian Muisca/Chibcha peoples (Reichel-Dolmatoff 1981: 29). Alongside other semi-precious minerals, gold, copper, sea shells, narcotics and woven goods, salt was a major feature of trade in Colombia (*ibid*: 30). These items seem to have shared a conceptual value associated with fertility, and for many of the region's peoples, salt symbolised sex, specifically semen, and had aphrodisiac qualities (*ibid*). It was also symbolically linked to lime – the fine white powder obtained from burning sea shells and used in coca chewing (*ibid*).

Equally insightful is the perception of naturally occurring salt sources in lowland Amazonia. Among the Nukak of the Colombian Amazon, salt-licks (*salados*) are regarded as 'doors' to the supernatural 'House of the Tapir'. Salt-licks attract prey animals, and the Nukak believe that:

> tapir, jaguar, deer, and other animals – whose tracks can be seen in the *salados* – use these locations to emerge from the underworld at night. In this sense, *salados* are sacred places rarely visited by Nukak. (Politis and Saunders 2002: 121)

Yet, while accumulations of salt are avoided by Nukak, other Amazonian groups such as the Yukuna will hunt near them after their shamans have negotiated hunting arrangements with the spirit owner of the *salado* (*ibid*). In this instance, accumulations of salt elicit different behaviours from generically similar hunter-gatherer groups.

While these are examples of the direct ingestion of minerals, a more subtle and symbolic association may also be discerned. Quartz, a mineral possessing important shamanic attributes (see below), is used in the form of small fragments inset into manioc graters throughout the Amazon (Mowat 1989: 21–24, 51–52). These graters are highly valued items, and are embedded in wide-ranging trade networks, as well as being involved in the physical and metaphysical transformation of raw and poisonous manioc (*Manihot esculenta*) into civilised (cooked and safe) food. The rubbing of manioc tubers over a quartz-encrusted grater may, in one sense, be seen as symbolically imbuing the food with the mineral's magical qualities. All food prepared in this way may perhaps be considered as having been 'mineralised'.

MOVING MINERALS: THE SIGNIFICANCE AND TRADE OF SACRED STONES

Amerindian worldview conceived the earth – its geographical features and geological components – as a spiritualised and philosophically coherent entity. From mountain peak to clay particle, active volcano to obsidian vein, landscapes were anthropomorphised, and considered pro-active in human affairs. In this sense, it is no surprise that small portable minerals should themselves possess a variety of cultural significances.

Scattered throughout the ethnohistoric and ethnographic sources are insights into the ways in which particular minerals were identified as embodying and representing social status and cultural identity in the Americas, especially in Inka Peru and Aztec Mexico. During the conquests of the Inka emperor Pachacutec, he encountered the Huamanchucu, who were said to have 'worshipped stones of various colours ... such as jasper. The idea was that different colours could only be produced in the same stone by the influence of some great god within it, and ... they kept such stones in their houses as idols' (Garcilaso de la Vega [1609] 1987: 342).

Inka belief also associated minerals with mythic ideas of the origins and nature of mummies. The Inka concept of *huanca* included tribute-offerings, but specifically represented the sacrificed body of an ancestral hero turned to stone – in other words, the 'mineral double' of a sacred mummy (Duviols, quoted in Sullivan 1988: 859). This is an Inka expression of the telluric cosmological preoccupation of Andean societies. Petrification of prominent figures from Inka mythology found its most elaborate form in the imperial capital at Cuzco – a city built literally and figuratively from stone on the site where Ayar Auca, the brother of the founding hero Manco Capac, was himself turned to stone.

Throughout the Andes, lowland Amazonia, Mesoamerica and the Caribbean Islands, many different kinds of minerals such as greenstones were highly valued and widely traded (Boomert 1987; Oliver 1989: 216–17). Mica was regarded as a precious cosmic substance in Mesoamerica, and at Hopewell in North America some 3,000 sheets of the material and 200 pounds of galena were excavated from a single mound (Moorehead 1922: 91–92). At the central Mexican metropolis of Teotihuacán, two layers of mica plates were discovered beneath a floor with mica debris above, suggesting to the excavator that the structure had originally been a glittering mica-plated building (Armillas 1944: 123–24). Even earlier, around 1150 BC at San José Mogote in Oaxaca, mica (together with jade and magnetite) had evidently been a symbol of high social status (Marcus and Flannery 1996: 93, 101–03).

In North America, minerals played important roles in sacred and secular life (see Charles *et al*, this volume). Among the Pueblo Indians of the south-western United States, white-flint arrowheads had the power to bestow clairvoyancy, and obsidian the power to cure disease (Gunnerson 1998: 242–43). In the north-eastern United States, the Huron-Wyandot 'White Lion' mythic being carried white stones which assured success in hunting and courtship and which are identified as quartz, white chert, flint and chalcedony (Hamell 1998: 270). The mythic associations of such minerals are present also in the white fire-dragon of the Seneca – a supernatural being who lived at the edge of the world 'in a cave beneath a great rock of red hot white chert or white flint' (*ibid*).

The variety of culture-specific valuations of minerals is apparent also in the prehistoric Caribbean. Throughout the region, the first pottery-making agriculturalist peoples of the Saladoid culture (500 BC – AD 300) invested considerable time and effort in acquiring and/or making polished stone beads and zoomorphic (often avian-shaped) pendants from carnelian, quartz, amethyst, nephrite and turquoise. Archaeological evidence from the Saladoid occupation of the island of Montserrat includes 29 varieties of minerals, none of which occurred naturally on the island. All appear to have been imported as part of an inter-regional trade in such luxury items that embraced the islands of Vieques, St Croix, Grenada and, ultimately, their geological source in South America (Watters and Scaglion 1994).

Much later, during Taíno (Arawak) times in the Greater Antilles, necklaces of polished multi-coloured stone beads known as *çibas* were part of the ritual attire of the elite (*caciques*) who first encountered Europeans (Oliver 2000: 205). Even at the time, Europeans observed that *çibas*, together with worked-shell, body painting, feathered headdresses and the gold-copper alloy known as *guanín*, had ritual significance. Together with other items, these objects – their shapes, designs, colours and media – are collectively known as *zemís*, omnipresent spirit helpers who send rain and sun, promote fertility and protect people (see Arrom 1999: 47).

Particularly well documented is the widespread occurrence of different varieties of quartz, and their embeddedness in Amerindian cultural life (eg Márquez and Eilson nd; Serra Puche and Solís Olguín 1994). In the Andes, rock crystals adorned Inka temples and were associated conceptually with pearls and rain (Mester 1990: 206, 216). More recently, rock crystals have been highly regarded for their curative properties throughout the Americas (Dow 1986: 108–10; Sharon 1978), and are the Amazonian Barasana shaman's most important power objects (Hugh-Jones 1979: 121). Also in the Amazon, the Desana consider a crystal's hexagonal shape to be an image of cosmic order, related to concepts of energy, transformation and fertility, with the crystal regarded as concentrated semen (Reichel-Dolmatoff 1975: 101–02). The sun is conceived as a huge rock crystal, and every earthly crystal a miniature sun (Reichel-Dolmatoff 1981: 23–24).

In Mexico, modern Nahuatl Indians see the sky as a huge crystal mirror, filled with the sparkle of the sun and stars (Sandstrom 1991: 238). In North America, the Chippewa-Ojibwa and Seneca Iroquois associate quartz crystals with shamanic curing, divination and the human soul (Hamell 1983: 13–14). In Navajo Athapaskan mythology, crystals symbolise 'clear seeing' and 'consciousness' (Witherspoon, quoted in Hamell 1986: 58).

A crystal world: the Kogi

Illustrating the complex and interwoven meanings of minerals for archaeological and recent Amerindian cultures in the same geographical/geological area is the wealth of information on northern Colombia published by Gerardo Reichel-Dolmatoff. Today, the Kogi inhabit parts of this region, and are considered the descendants of the prehistoric and contact-period Tairona.

In Kogi worldview, their entire Andean homeland of the Sierra Nevada is conceived as a huge rock crystal, a petrified world where each mountain is a temple and snow peaks are gleaming white crystals (Reichel-Dolmatoff 1981: 28). Even cosmogony is seen through the lens of ethnomineralogy according to a creation myth of the nearby Muisca/Chibcha people. Here, a chief's daughter became pregnant after exposing her body to the fertilising light of the dawn sun – subsequently giving birth to a huge emerald which contained a future great chief (Reichel-Dolmatoff 1981: 29). The Kogi themselves keep rock crystals as treasured heirlooms, and these objects are associated with a priestly lineage known as *kurcha* or 'Owners and Keepers of the Rock Crystal' (*ibid*).

Other minerals are also highly valued, including 'carnelians, agates, with beautiful veinings, chalcedonies, and many others' (*ibid*: 26). These minerals are sometimes found in caches associated with prehistoric burials in the region, and it seems that the pre-Columbian Tairona were expert in making and polishing them into beads (*ibid*: 27). Today, female Kogi still wear bead necklaces, though these are now a mixture of 'precious archaeological specimens ... combined with modern trade beads of glass or porcelain' (*ibid*). Kogi priests identify at least 50 kinds of mineral beads, used for curing, offerings, and the protection of their owners against spiritual aggression. Some are 'damaged' by a piece being chipped off, ground into powder, and later ritually blown as part of an offering ceremony.

When the Kogi build a new house or temple, they bury caches of necklace beads placed in a clay vessel in the centre of the structure. These minerals (perforated for women, unperforated for men) symbolically represent Kogi society on a number of levels. If the new building is a house, then the number of beads interred corresponds to the number of people who will inhabit it; if it is a temple, then the number of beads will correspond to all the members of the initiated men's families (*ibid*).

At a deeper level, every Kogi is associated through their lineage with certain minerals and colours, the lineage itself regarded as the owner of these minerals and their symbolic attributes. Such ideas are linked in turn to wider spiritual beliefs, as each lineage and mineral corresponds to an aspect of the natural world such as a plant, animal or cardinal direction, each of which has its own 'mother' and 'father' who call for offerings of the appropriate kinds of polished-stone beads. Here, minerals may be considered food for the spirits, the ritual offering of which protects and fertilises the natural food, animal or plant that corresponds to it (*ibid*).

Mesoamerica: greenstones and obsidian

Illustrating these ideas in more detail within one culture area is possible for Mesoamerica, where there exists a rich ethnohistorical record – such as the 16th-century *Florentine Codex* of Bernardino de Sahagún. In Book 11, *Earthly Things* (223, 229–30), Sahagún makes constant reference to the translucence, preciousness, iridescent colours and light-giving qualities of minerals, thereby illustrating the distinctive Mesoamerican Aztec view of the nature and place of minerals in the Postclassic Central Mexican worldview.

Sahagún records the physical attributes of minerals and the ways these were interpreted according to the Aztec experience of everyday life, their animistic religion, and the resulting perceptions of the ways in which the natural and supernatural worlds mingle endlessly in physical objects. He observes how 'fine turquoise' is called *teuxiujtl*, whose name is formed by adding the prefix *teotl* (god) to *xiuitl* (turquoise) and thereby denoting that it belongs to the deities and that it smokes and appears like the beautiful cotinga bird (Book 11: 224) (see **Figure 7.1 colour section**). Opal, whose name *vitzitziltetl* comes from *uitzitzilin* (hummingbird) and *tetl* (stone), is so called because its shimmering appearance is like the feathers of the hummingbird (*ibid*: 229–30). It was regarded as precious and esteemed, and radiated a glow akin to a group of fireflies (*ibid*).

Amber (*apoçonalli*) took its name from water (*atl*) and bubble (*poçonalli*), and at dawn 'when the sun rises, it appears just like a foam which the sun's rays penetrate ... it is as if little sparks continually fly from it, as if a flame stood within it' (*ibid*: 225). Mica, *metzcujtatl*, was believed to be excreted by the moon – it symbolised cosmic forces, and radiated a soft, buoyant light (*ibid*: 235). These excerpts from Sahagún's list of minerals provide valuable insights into Aztec ethnomineralogy which become sharply focused in his description of rock crystal, 'the tool of conjurors, in the lucid depth of which they discern the past, the future, and all secret things' (Seler 1991: 281).

In Aztec natural philosophy, each mineral was not only a raw material for future artefacts, but also possessed spiritual and ritual qualities that articulated a worldview where every kind of matter was infused with spiritual power. Children who died before being weaned were regarded as untainted by everyday life – so pure in fact that they are 'green stones, they become precious turquoise, they become bracelets' (López Austin 1988: 314). The power of such minerals to symbolise life was extended by the belief that they could energise and animate the less powerful kinds of stone used to make large statues. Weisman (1950: 13) observes that such figures often had:

> a disk of semiprecious stone, like obsidian or jadeite, set into the torso: this disk represents the godhead, the soul, giving life to the graven image.

Practices such as the insetting of powerful stones survived the Spanish military and spiritual conquest of Mexico, as culturally hybridised material forms (García Granados 1940–42). In discussing the conflation of Aztec female deities and the Christian Virgin, Calloway (1989: 237–38) notes that at the monastery of San Agustin at Acolman, a statue of 'the colonial Virgin wears a disk which ... may represent "that inset precious stone" signifying the deity's heart, a purely indigenous characteristic'.

Beyond the Aztec political frontier, similar ideas were shared by contemporary Mesoamerican peoples. Among the Huichol of the western Mexican state of Jalisco, clouds were associated with crystals and human souls. In particular, the physical appearance of chalcedony symbolised clouds and was thus appropriate to be made into festishes for rainmaking (Seler 1993a: 195). In the Tarascan kingdom to the west of the Aztec empire, kings and warriors wore crystal lip-plugs, thereby associating themselves with supernatural power by incorporating minerals directly into their bodies (*ibid*: 31).

Arguably the two most important minerals that played a central role in Mesoamerican ritual and everyday life for at least 3,000 years were greenstones and obsidian. According to Sahagún (Book 11: 221–22), greenstones connoted fertility and glistening preciousness, and signalled their presence at dawn by emitting smoke, magically imparting greenness to the flora in their vicinity. This life-giving quality was signified by placing the greenstone called *chalchihuitl* in the mouth of a deceased emperor (López Austin 1988: 326–27).

The complexities of mineralogy in Aztec worldview are indicated here by the sophisticated imagery surrounding this stone: the Aztec hieroglyphic combination of *chalchihuitl* (green jewel) in a stream of water (*atl*) signifies 'precious water', another name for sacrifical human blood (Seler 1993b: 277). In Aztec cosmology and myth, human blood was deemed a crucial offering to the gods, both as a debt-repayment for bringing the world into being, and to energise the Sun God Tonatiuh during his nocturnal journey through the underworld.

The dark volcanic glass obsidian similarly possessed polysemic qualities in the Aztec worldview. In the Florentine Codex, obsidian is named *itztli*, and is identified as the material from which razors and awls for sacrificial bloodletting are made (Sahagún, Book 11: 226–27). In Aztec metaphysics, there appears to have been an association of 'obsidian, its sources, and local inhabitants, in ways which enabled the material to become a metaphorical embodiment – a reification – of cosmic and earthly identity' (Saunders 2001: 222). This was the case at

Ytztepexic, a town which took its name from a nearby outcrop of obsidian, where a great obsidian knife was worshipped as a deified 'talking stone' in the community's temple (Heyden 1981: 25).

The cultural value of obsidian was also ambiguous, as it possessed positive and negative uses. When ground down, obsidian powder was put into the eyes of those suffering from cataracts, an Aztec practice that continued at least up until the 16th century (Clark 1989: 315). Obsidian dust was also believed to help heal fresh wounds and rheumatism, yet apparently was also used to kill people when mixed with their drink (*ibid*: 316).

Nowhere was obsidian more embedded in Mesoamerican symbolic thought than in its apotheosis as Tezcatlipoca, the supreme Aztec deity whose name meant 'Lord of the Smoking Mirror' (Nicholson 1971: 412). The Spanish chronicler Diego Durán (1971: 98) observed that the god's temple image was made of a lustrous stone, black as jet, from which sharp blades and knives were made – ie obsidian (Heyden 1988: 222).

Obsidian's peerless utility in a world without metal tools in part generated an enduring Mesoamerican aesthetic that saw those who controlled obsidian sources and made obsidian blades as connected to cosmic forces. Subsequent acts of obsidian use were endowed with a potency and significance, particularly in rituals of sacrifice and bloodletting.

Such were the deep semantic connections between obsidian, belief and ritual that the phrase *tentli-itztli* meant to 'sacrifice and cut one's lips before the idols' (Campbell 1985: 139) – a linguistic identification of mineral with ritual act. Obsidian's role as transformed matter bridged physical and spiritual worlds, and possessed mythological and religious dimensions (see Saunders 2002: 222–24) (**Figure 7.2**).

Figure 7.2: 16th century atrial cross with obsidian disk insert. Michacan, Mexico (photo Nicholas J Saunders).

Light and magic, the music of minerals

The multi-layered worldview within which minerals are centrally located extends also to sensorial domains that in Western culture are often subordinated to sight. The aural dimension of sound, particularly music, is notable in this respect. Not only is music an integral part of Amerindian natural and moral philosophies, as recognised and explored by Basso (1985) in *A Musical View of the Universe*, but, more specifically, it exhibits a mineralogical dimension. Musical instruments incorporating minerals add percussive sound to ceremonial activities, and reinforce the emotional intensity of such events by adding unexpected sensorial stimulation.

Music in Amerindian life was, and remains, an integral part of worldview. Singing and chanting maintains ideas of community, and praises ancestors and the spirits of the forest. Amerindian singing, chanting, whistling, percussion and dancing possess a numinous quality, connecting social life to the spiritual breath that permeates the universe. Human breath has magical power to summon spirits, and shamans blow protective and purifying spells over people and food alike (Butt-Colson 1977; Hugh-Jones 1979: 91–100). In pre-Columbian times, many musical instruments were made from fired clay into pottery flutes, trumpets, whistles, drums, pan pipes and rattles (Cabello and Martínez 1988; Olsen 2002), suggesting perhaps a connection of transformations between clay, fire and ceramic (mentioned above) with ideas of the inherent spirituality of music.

A recent ethnographic example illustrates connections that may once have been commonplace amongst pre-Columbian societies. The shamans of the Warao of coastal Venezuela possess a percussion instrument known as the *hebu mataro* that embodies and activates complex linkages between shamanism, music, ethnomineralogy and the laws of physics.

The *hebu mataro* is a calabash rattle, some 70 cm long with a wooden handle, that is decorated with red and yellow feathers of the *cotora* parrot. Inside the rattle are small quartz fragments that glow and spark when the rattle is shaken during nocturnal curing trances. These pieces of quartz represent crystalline spirits in Warao thinking, demonstrating the association between physical and metaphysical worlds. The importance of minerals, human breath, spirituality and music are shown when:

> the shaman blows tobacco smoke ... over scores of small quartz crystals and places these one by one into the hollowed-out gourd, invoking for each a particular ancestral spirit believed by the Warao to be embodied in the stone. These quartz crystals are his familiar spirits who assist him in curing and other ritual endeavours; ... he refers to them as his 'family'. (Wilbert 1977: 58)

Warao shamans use the rattle to cure illness, the sparking light of the quartz helping to convince the patient of the shaman's powers.

Almost identical is the gourd rattle used by the modern Moche shaman Eduardo Calderón on Peru's north coast. It too accompanies whistling and songs designed to attract guardian spirits. Calderón's rattle, variously called a *chungana* or *maraca,* has small pieces of lapis lazuli, flint and turquoise inside, as well as seeds. When the rattle is shaken in darkness, it emits sparks (José Gushiken quoted in Olsen 2002: 158).

These two examples, alongside similar ones from North America, where sparking rattles summon the white buffalo (*ibid*: 159), demonstrate how the physical properties of certain minerals articulate worldview through ritual.

From a Western scientific perspective, the phenomenon of quartz luminescence is itself only partly understood. The basis seems to be that the molecular composition of quartz's crystalline structure is such that when two pieces are brought together by impact or friction, they emit a particular kind of light known as triboluminescence (Oldale 2003: 1–2). This is thought to be related to the piezoelectric nature of quartz (*ibid*).

Piezoelectricity is produced when mechanical stress causes electrons in quartz to flow and produce an electric current that ionises gases trapped in the quartz in the same way that electricity passing between earth and clouds produces a lightning bolt (*ibid*). Here, Western science and Amerindian metaphysics meet, as Amazonian Desana shamans collect quartz crystals that they consider to have been scattered across the landscape by lightning (Reichel-Dolmatoff 1975: 10b2). Human bone is also piezoelectric, an attribute that promotes healing in fractures – a coincidental and ironic parallel to the Warao shaman's shaking his *hebu mataro* rattle and producing a sparking piezoelectric charge in rituals designed to cure illness and heal wounds.

CONCLUSIONS

Geology, geography and landscape are Western concepts. As such, their uncritical application to prehistoric and traditional non-Western societies often obscures more than it reveals. By dismantling a sophisticated and holistic Amerindian worldview, and forcing its parts into the structures of Western categories, indigenous subtleties of meaning and significance are elided rather than understood, and may disappear from view completely. This is true despite the many interesting parallels between our own (Western) folklore of minerals (eg Kunz 1971) and current indigenous Amerindian beliefs.

In pre-Columbian and Native America, worldviews were created, structured and maintained by symbolic analogical reasoning rather than Western notions of rationality and Linnaean classification. Adhering to their own principles of categorisation, Amerindian worlds were conceived as animated by spirits and ancestors, and – problematically for us – as being actively transformational in appearance and spiritual essence. In order to understand the cultural life of minerals in such worlds, it is necessary to consider a wide range of natural phenomena and social action that together constituted the meaningful frameworks for their valuations and uses.

We can see from the wealth of archaeological and ethnographic data outlined above that the meaning of minerals was (and in places remains) inextricably bound to natural and moral philosophies that regarded the physical world as a living entity rather than an inanimate and inert background for human activity. In this way, minerals are integral parts of larger entities, such as mountains, volcanoes, lakes and valleys, and retain parts of the (often mythological) significance accorded to these landscape features. Minerals may also partake of the cultural values ascribed to wider, less bounded environmental features, such as rivers, floodwaters, rain, ice and lightning, in a bewildering array of meanings which are coherent only (or mainly) for the culture concerned. The kaleidoscopic relativity of cultural meanings may associate some minerals with water, blood, social status, ethnic and spiritual identity, and be used metaphorically to signify certain identifications and qualities pertaining to the dead. Characterised by certain formations of minerals, some areas may be identified with particular food or medicinal plants, and animals, in ways that draw attention to the interchangeability of outward form that is such a prominent feature of Amerindian worldview.

The anthropological study of landscape should not privilege only the political and ideological nature and manipulation of place. It must also acknowledge that at one level a landscape is a blanket term for a physical and cultural entity that is constituted of many mineral (and other) components, each of which possesses its own meanings and significances. Each Amerindian culture engaged with its physical surroundings in specific and historically contingent ways, emphasising the sensorial, spiritual, economic and ideological qualities of the minerals that made up its physical surroundings, and which in turn were used to construct a cultural environment. Yet, lying beneath these differing cultural interpretations of what minerals meant for a society at a particular time was a pan-Amerindian engagement with the natural world – a shared appraisal of physical surroundings and the elements that constituted it, whether minerals, flora, fauna, or meteorological phenomena.

In the Americas, and probably elsewhere, the ethnohistoric and ethnographic records provide a window into the prehistoric past that, while not without interpretive problems, nevertheless hints at the complex interconnections and transformations between what Western societies regard as irreconcilably different kinds of matter. In the case of minerals, we are dealing literally with the fundamental elements of the archaeological record whose commonplace nature for us can easily disguise possible significances for those who produced it. As we sieve the soil we excavate, we should consider the earth as a landscape of minerals – each one of which may have embodied a part of a meaningful cultural landscape in the past.

The associations of minerals with cosmology, identity, spirituality and materiality offer significant opportunities for anthropologically informed archaeological research. The richness of evidence available for the Americas (and only hinted at here) may not always be available for other times and places. Nevertheless, the multi-dimensional and multi-sensorial incorporation of minerals into Amerindian cultures at every level is an indication of how worthwhile such investigations might be.

References

Armillas, P (1944) 'Exploraciones recientes en Teotihuacán', *Cuadernos Americanos* XVI (4), 121–36

Arnold, DE (1993) *Ecology and Ceramic Production in an Andean Community*, Cambridge: CUP

Arrom, JJ (ed) (1999) *An Account of the Antiquities of the Indians. Fray Ramón Pané*, Durham, NC: Duke University Press

Basso, EB (1985) *A Musical View of the Universe*, Philadelphia: University of Pennsylvania Press

Bastien, JW (1985) *Mountain of the Condor: Metaphor and Ritual in an Andean Ayllu*, Prospect Heights, Ill: Waveland

Berthelot, J (1986) 'The extraction of precious metals at the time of the Inka', in Murra, JV, Wachtel, N and Revel, J (eds), *Anthropological History of Andean Polities*, Cambridge: CUP, pp 69–88

Boomert, A (1987) 'Gifts of the Amazons: "greenstone" pendants and beads as items of ceremonial exchange in Amazonia' 67 *Antropologia* 33–54

Broda, J (1987) 'Templo Mayor as ritual space', in Broda, J, Carrasco, D and Matos Moctezuma, E, *The Great Temple of Tenochtitlan: Center and Periphery in the Aztec World*, Berkeley, CA: University of California Press, pp 61–123

Burkhart, L (1992) '"Flowery heaven": the aesthetic of paradise in Nahuatl devotional literature' 21 *Res: Anthropology & Aesthetics* 89–109

Butt-Colson, A (1977) 'The Akawaio Shaman', in Basso, EB (ed), *Carib-Speaking Indians: Culture, Society and Language*, Anthropological Papers of The University of Arizona 28, pp 43–65

Cabello, P and Martínez, C (1988) *Música y Arqueología en América Precolombina: Estudio de una colección de instrumentos y escenas musicales*, Oxford: BAR International Series 450

Calloway, CH (1989) *The Church of Nuestra Señora de la Soledad in Oaxaca, Mexico*, PhD thesis, University of Maryland, Ann Arbor: University Microfilms

Campbell, RJ (1985) *A Morphological Dictionary of Classical Nahuatl: A Morpheme Index to the Vocabulario en lengua mexicana y castellana of Fray Alonso de Molina*, Madison: Hispanic Seminary of Medieval Studies

Clark, JE (1989) 'Obsidian: the primary sources', in Gaxiola, G and Clark, JE (eds), *La Obsidiana en Mesoamérica*, Mexico: Instituto Nacional de Antropología e Historia, pp 299–319

Dow, J (1986) *The Shaman's Touch: Otomí Indian Symbolic Healing*, Salt Lake City: University of Utah Press

Durán, Fray Diego de (1971) *Book of the Gods and Rites and the Ancient Calendar*, Heyden, D and Horcasitas, F (trans), Norman: University of Oklahoma Press

García Granados, R (1940–42) 'Reminiscencias idolatricas en monumentos coloniales' 2 *Anales del Instituto de Investigaciones Estéticas* 54–56

Garcilaso de la Vega [1609] (1987) *Royal Commentaries of the Incas and General History of Peru*, Austin: University of Texas Press

Gasparini, G and Margolies, L (1980) *Inca Architecture*, Bloomington: Indiana University Press

Gell, A (1998) *Art and Agency: An Anthropological Theory*, Oxford: Clarendon

Gibson, JJ (1986) *The Ecological Approach to Visual Perception*, Hillsdale, NJ: Lawrence Erlbaum Associates

Gunnerson, JH (1998) 'Mountain lions and pueblo shrines in the American southwest', in Saunders, NJ (ed), *Icons of Power: Feline Symbolism in the Americas*, London: Routledge, pp 228–57

Hamell, GR (1983) 'Trading in metaphors: the magic of glass beads. Another perspective upon Indian-European contact in northeastern North America', *Proceedings of the 1982 Glass Bead Conference*, CF Hayes (ed), Research Records 15, Rochester, NY: Rochester Museum and Science Center, pp 5–28

Hamell, GR (1986) *Wampum among the Northern Iroquoians: A Proposed Semantics of Color, Ritual, and Material Culture*, Paper prepared for the 1989 Shell Bead Conference, Rochester Museum and Science Center, New York

Hamell, GR (1998) 'Long-tail: the panther in Huron-Wyandot and Seneca myth, ritual, and material culture', in Saunders, NJ (ed), *Icons of Power: Feline Symbolism in the Americas*, London: Routledge, pp 258–91

Helms, MW (1986) 'Art styles and interaction spheres in Central America and the Caribbean: polished black wood in the Greater Antilles' 12(1) *Journal of Latin American Lore* 25–43

Heyden, D (1981) 'Caves, gods, and myths: world-view and planning in Teotihuacan', in Benson, EP (ed), *Mesoamerican Sites and World-Views*, Washington, DC: Dumbarton Oaks, p 140

Heyden, D (1988) 'Black magic: obsidian in symbolism and metaphor', in Josserand, JK and Dakin, K (eds), *Smoke and Mist: Mesoamerican Studies in Memory of Thelma D Sullivan*, Oxford: BAR International Series 402, pp 217–36

Hosler, D (1994) *The Sounds and Colors of Power: The Sacred Metallurgical Technology of Ancient West Mexico*, Cambridge, MA: MIT Press

Houston, S and Taube, K (2000) 'An archaeology of the senses: perception and cultural expression in ancient Mesoamerica' 10(2) *Cambridge Archaeological Journal* 261–94

Hugh-Jones, S (1979) *The Palm and the Pleiades: Initiation and Cosmology in Northwest Amazonia*, Cambridge: CUP

Hyslop, J (1990) *Inca Settlement Planning*, Austin: University of Texas Press

Karsten, R (1968) *The Civilization of the South American Indians, With Special Reference to Magic and Religion*, London: Dawsons of Pall Mall

Kirchoff, P (1948) 'The Otomac', in Steward, JH (ed), *Handbook of South American Indians 4*, Washington, DC: Smithsonian Institution, pp 439–44

Kunz, GF (1971) *The Curious Lore of Precious Stones*, New York: Dover

Lechtman, H (1993) 'Technologies of power: the Andean case', in Henderson, JS and Netherly, PJ (eds), *Configurations of Power*, Ithaca: Cornell University Press, pp 244–80

López Austin, A (1988) *The Human Body and Ideology: Concepts of the Ancient Nahuas*, 2 vols, Ortiz de Montellano, T and Ortiz de Montellano, B (trans), Salt Lake City: University of Utah Press

Lothrop, SK (1946) 'Indians of the Paraná and La Plata Littoral', in Steward, JH (ed), *Handbook of South American Indians 1*, Washington, DC: Smithsonian Institution, pp 177–90

Lunt, SW (1988) 'The manufacture of the Inca Aryballus', in Saunders, NJ and de Montmollin, O (eds), *Recent Studies in Pre-Columbian Archaeology 2*, Oxford: BAR International Series, pp 489–511

Marcus, J and Flannery, KV (1996) *Zapotec Civilization: How Urban Society Evolved in Mexico's Oaxaca Valley,* London: Thames and Hudson

Márquez, JD and Eilson, JE (nd) *Pre-Columbian Quartz Sculptures,* Caracas: Graficas Armitano CA

Mester, AM (1990) *The Pearl Divers of Los Frailes: Archaeological and Ethnohistorical Explorations of Sumptuary Good Trade and Cosmology in the North and Central Andes,* PhD thesis, University of Illinois at Urbana-Champaign, Ann Arbor: University Microfilms

Métraux, A (1948) 'The hunting and gathering tribes of the Rio Negro Basin', in Steward, JH (ed), *Handbook of South American Indians 3,* Washington, DC: Smithsonian Institution, pp 861–67

Moorehead, WK (1922) *The Hopewell Mound Group of Ohio,* Field Museum of Natural History Publication 211, Anthropological Series VI (5), Chicago: Field Museum of Natural History

Mowat, L (1989) *Cassava and Chicha: Bread and Beer of the Amazonian Indians,* Princes Risborough: Shire

Murra, JV (1975) *Formaciones económicas y políticas del mundo andino,* Lima: Instituto de estudios Peruanos

Nicholson, HB (1971) 'Religion in pre-Hispanic Central Mexico', in Eckholm, G and Bernal, I (eds), *Handbook of Middle American Indians, 10, Archaeology of Northern Mesoamerica 1,* Austin: University of Texas Press, pp 395–445

Niederberger, C (2000) 'Ranked societies, iconographic complexity, and economic wealth in the Basin of Mexico toward 1200 BC', in Clark, JE and Pye, ME (eds), *Olmec Art and Archaeology in Mesoamerica,* New Haven: Yale University Press, pp 169–91

Oldale, RN (2003) 'Cape Cod's "magic" quartz pebbles', online at http://woodshole.er.usgs.gov/staffpages/boldale/pebbles/index/html

Oliver, JR (1989) *The Archaeological, Linguistic and Ethnological Evidence for the Expansion of Arawakan into Northwestern Venezuela and Northeastern Colombia,* unpublished PhD thesis, University of Illinois at Urbana-Champaign

Oliver, JR (2000) 'Gold symbolism among Caribbean chiefdoms: of feathers, çibas, and *guanín* power among Taino elites', in McEwan, C (ed), *Precolumbian Gold: Technology, Style and Iconography,* London: British Museum Press, pp 196–219

Olsen, DA (2002) *Music of El Dorado: The Ethnomusicology of Ancient South American Cultures,* Gainesville, FL: University Press of Florida

Politis, GG and Saunders, NJ (2002) 'Archaeological correlates of ideological activity: food taboos and spirit-animals in an Amazonian hunter-gatherer society', in Miracle, P and Milner, N (eds), *Consuming Passions and Patterns of Consumption,* Cambridge: McDonald Institute Monographs, pp 113–30

Reichel-Dolmatoff, G (1975) *The Shaman and the Jaguar: A Study of Narcotic Drugs Among the Indians of Colombia,* Philadelphia: Temple University Press

Reichel-Dolmatoff, G (1981) 'Things of beauty replete with meaning: metals and crystals in Colombian Indian cosmology', in *Sweat of the Sun, Tears of the Moon: Gold and Emerald Treasures in Colombia,* Los Angeles: Natural History Museum of Los Angeles, pp 17–33

Reinhard, J (1988) 'The Nazca lines, water and mountains: an ethnoarchaeological study', in Saunders, NJ and de Montmollin, O (eds), *Recent Studies in Pre-Columbian Archaeology,* Oxford: BAR International Series 421, pp 363–414

Reinhard, J (1991) *Machu Picchu: The Sacred Center,* Lima: Nuevas Imágenes

Reinhard, J (1992) 'An archaeological investigation of Inca ceremonial platforms on the volcano Copiapo, central Chile', in Saunders, NJ (ed), *Ancient America: Contributions to New World Archaeology*, Oxford: Oxbow, pp 145–72

Reinhard, J (1996) 'Peru's ice maidens' 189(6) *National Geographic* 62–81

Sahagún, B de (1950–78) *Florentine Codex: General History of the Things of New Spain*, Anderson, AJO and Dibble, CE (trans and ed), 13 vols, Santa Fé: School of American Research and the University of Utah

Sandstrom, AR (1991) *Corn is our Blood: Culture and Ethnic Identity in a Contemporary Aztec Indian Village*, Norman: University of Oklahoma Press

Saunders, NJ (1998) 'Stealers of light, traders in brilliance: Amerindian metaphysics in the mirror of conquest' 33(1) *RES: Anthropology and Aesthetics* 225–52

Saunders, NJ (1999) 'Biographies of brilliance: pearls, transformations of matter and being, c AD 1492' 31(2) *World Archaeology* 243–57

Saunders, NJ (2001) 'A dark light: reflections on obsidian in Mesoamerica' 33(2) *World Archaeology* 220–36

Saunders, NJ (2002) 'Mirror or portable altar', in Matos Moctezuma, E and Solís Olguín, F (eds), *Aztecs*, London: Royal Academy Publications, p 84

Saunders, NJ (2003) '"Catching the light": technologies of power and enchantment in pre-Columbian goldworking', in Quilter, J and Hoopes, JW (eds), *Gold and Power: In Ancient Costa Rica, Panama, and Colombia*, Washington, DC: Dumbarton Oaks, pp 15–47

Seler, E (1991) 'The jewellry craft of the ancient Mexicans and their art of working in stone and making feather ornaments', in Thompson, JES and Richardson, FB (eds), *Eduard Seler, Collected Works in Mesoamerican Linguistics and Archaeology*, Vol II, Culver City, CA: Labyrinthos, pp 270–95

Seler, E (1993a) 'The Huichol Indians of the State of Jalisco in Mexico', in Thompson, JES and Richardson, FB (eds), *Eduard Seler, Collected Works in Mesoamerican Linguistics and Archaeology*, Vol IV, Culver City, CA: Labyrinthos, pp 179–97

Seler, E (1993b) 'The wall sculptures in the Temple of the Pulque God at Tepotztlan', in Thompson, JES and Richardson, FB (eds), *Eduard Seler, Collected Works in Mesoamerican Linguistics and Archaeology*, Vol IV, Culver City, CA: Labyrinthos, pp 266–80

Serra Puche, MC and Solís Olguín, F (1994) *Cristales y Obsidiana Prehispánicos*, México: Siglo XXI editores

Sharon, D (1978) *Wizard of the Four Winds: A Shaman's Story*, London: Free Press

Sullivan, LE (1988) *Icanchu's Drum: An Orientation to Meaning in South American Religions*, London: Macmillan

Watters, DR and Scaglion, R (1994) 'Beads and pendants from Trants, Montserrat: implications for the pre-historic lapidary industry of the Caribbean' 63 *Annals of the Carnegie Museum* 215–37

Weisman, EW (1950) *Mexico in Sculpture 1521–1821*, Cambridge, MA: Harvard UP

Wilbert, J (1977) 'The Calabash of the Ruffled Feathers', in Brodzky, AT, Daneswich, R and Johnson, N (eds), *Stones, Bones and Skin: Ritual and Shamanic Art*, Toronto: Artscanada, Society for Art Publications, pp 58–61

– 8 –

AN AXE TO GRIND

SYMBOLIC CONSIDERATIONS OF STONE AXE USE IN ANCIENT AUSTRALIA

Adam Brumm

INTRODUCTION

Innovative multidisciplinary research has the potential to provide valuable insights into symbolic aspects of prehistoric lithic procurement, use and other stone tool-related activities in ancient Australia. Worthy attempts towards this end have already been made in Australia in recent decades (eg Evans and Jones 1997; Fullagar and Head 1999; Graham and Thorley 1996; Jones 1990; Jones and White 1988; Moore 2003; Paton 1994; Taçon 1991). Elsewhere in the world, archaeologists emphasise the need to move beyond gross functionalist interpretations of lithics that divorce past human interaction with rocks and minerals from wider social, symbolic, cognitive, religious and phenomenological contexts (eg Cooney 1998; Dobres 2000; Edmonds 1995; Gero 1989; Hampton 1999; Kenoyer 1993; Postgate 1997; Sassaman 1997; Saunders 2001; Sinclair 1995; Stout 2002; Warburton and Duke 1995 among others; and see also chapters in the present volume).

Aside from casual remarks or off-the-cuff references to the possible ceremonial functions of unusual or elaborately manufactured specimens (eg Hayden 1979: 109; McCarthy 1976: 51), no serious archaeological attempts have been made to explore non-functional aspects of the production and use of the ubiquitous Australian stone axe (but see Davidson *et al* forthcoming; McBryde 1978, 1984, 1986, 1987, 1997 for research into social contexts of axe exchange). This represents a deficiency in archaeological understandings of both the tool type and Australian stone industries in general. As the products of social beings, like all material technologies employed by humans (Dobres 2000; Graves-Brown 1995, 2000), stone axes may have been embedded in multifarious worlds of socially constructed meaning, symbolism and power. Without a more detailed consideration of such meaningful aspects of prehistoric stone tools, we are seeing only a part of the picture.

In this chapter, I will consider evidence from ethnography, archaeology and rock art studies in Australia that together indicate that both stone and stone axes have likely played more than straightforward utilitarian roles in Aboriginal society from a very early stage in Australia's prehistory. As will be discussed, stone axe imagery in ancient rock art is frequently associated with subject matter of arguably supernatural content. This suggests that stone axes were symbolically meaningful objects for Aboriginal people in the ancient past, just as they were in the recent ethnographic past (see Sharp 1952), and were not simply profane utility equipment. Despite warranted cautions by some (eg Davidson 1999; Morphy 1999), rock art has the potential to provide tangible glimpses into the nature of past Aboriginal cultural ontologies and the antiquity of contemporary belief systems (David 2002; see also Chippindale *et al* 2000; Taçon 1991; Taçon and Chippindale 2001, 2002; Taçon *et al* 1996).

Of course, we can probably never know the precise 'meanings' behind ancient rock art imagery (Davidson 1999). But by considering the evocative ways in which some Aboriginal image-makers chose to depict stone axes in their art, we can gain valuable insights into elements of this tool's ancient symbolic associations, and the meaningfulness of human interactions with the materiality of stone in Australia. At the very least, enigmatic rock art portrayals of stone axes tend to imply that at various points in prehistory, Aboriginal peoples regarded these tools as more than meaningless utility implements, and stone as more than inert mineralogical matter, interpretations that are supported by the ethnographic and archaeological records. As Morwood (2002: 152) comments on representational images in Australian rock art:

> ... they are cultural transformations of reality ... but they are also selective. They do not portray a random sample of all possible objects; instead they tend to emphasize some subjects at the expense of others ... [T]he choice of subjects and ... the way they are portrayed, can tell us much about past values, cognitive systems, beliefs and ideologies.

THE AUSTRALIAN STONE AXE

At the time of European invasion, Aboriginal groups in many parts of Australia employed a stone implement known to archaeologists today as an edge-ground stone axe, hatchet or tomahawk (see **Figure 8.1**). This was a composite tool generally made by grinding an edge on a flaked hardstone blank and attaching a flexible withy handle using adhesive resin and binding (Dickson 1981). Early in colonial history, Europeans observed Aboriginal men and women using this tool to chop particular foods like 'sugarbag' (the honey of the stingless Trigona wild bee), possums and grubs from trees, as well as to perform a seemingly endless range of other woodworking and foraging tasks, such as splitting wood, removing bark from trees, fashioning canoes and other wooden artefacts, and butchering game (Smyth 1876; Roth 1897, 1904).

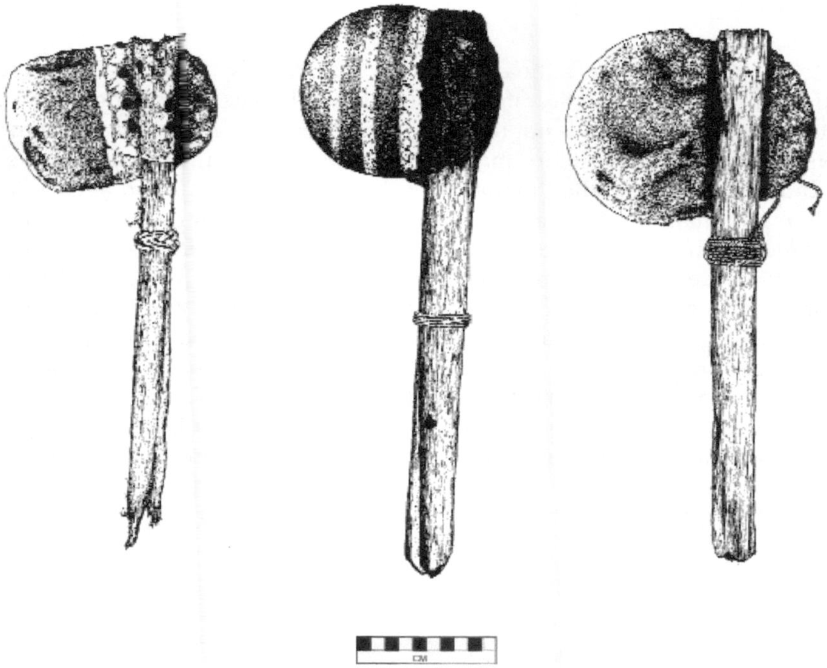

Figure 8.1: Stone axes collected from various parts of northern Australia in the late 1800s (courtesy of the Queensland Museum, illustrated by Adam Brumm).

The first archaeological evidence for the ancient use of stone axe technology in Australia came from Schrire's excavations at the Malangangerr and Nawamoyn rockshelters in western Arnhem Land in the 1960s, establishing a surprising Pleistocene antiquity of at least 20,000 years for the tools (Schrire 1982). More recent archaeological investigations in other parts of northern Australia have extended this time-frame, with axe material from excavated deposits dating to 28000 BP in the Kimberley (O'Connor 1999), 30000 BP in north-west Arnhem Land (Jones and Johnson 1985: 217–18) and 32000 BP in eastern Cape York (Morwood and Trezise 1989: 77).

These remarkably early contexts predate the development of edge-ground stone axe technologies elsewhere in the world, although Pleistocene axes have been recorded in Japan at 30000 BP (Imamura 1996: 25). Stone axes first appear in regions outside of northern Australia, with the exceptions of Tasmania and most of south-western Australia, only during the mid to late Holocene periods (Morwood and Hobbs 1995). In the desert and arid regions of central Australia, no stone axes have been recorded in deposits older than 1,000 years (*ibid*). There is currently no satisfying explanation as to why such an apparently useful technology was not in use in southern and central Australia until only relatively recent times (Morwood 2002: 17; Mulvaney and Kamminga 1999: 221).

THE POWER OF STONE

Western expressions like 'heart of stone', 'a stony glare' and 'stone cold' emphasise the lifeless quality of rocks and minerals. The perception of stone as inert geological substance is not universal, however, even in Western thought (eg Goldsworthy 1994). Indeed, for many cultures in the world, stone is perceived to be vibrantly alive, embodied matter with thoughts, intentions and agency much like living human beings. Maori people of New Zealand, for example, consider pounamou or nephrite to be 'ensouled', and individual pieces of the lustrous green stone are gendered and attributed social identity (Harvey 2000: 160). Amongst Yolngu people of eastern Arnhem Land, quartzite from Ngilipitji quarry (see **Figure 8.2**) is thought to grow in the ground like a living thing, and can also be 'pregnant' with 'baby stones' or 'eggs' (Jones 1990: 27; Jones and White 1988; see also Stout 2002 for Langda perceptions of 'living' stone). Belief in the life-force of inanimate stone objects is a common attribute of Indigenous shamanistic cultures (eg Bernstein 1997), which tend to perceive a fluidity rather

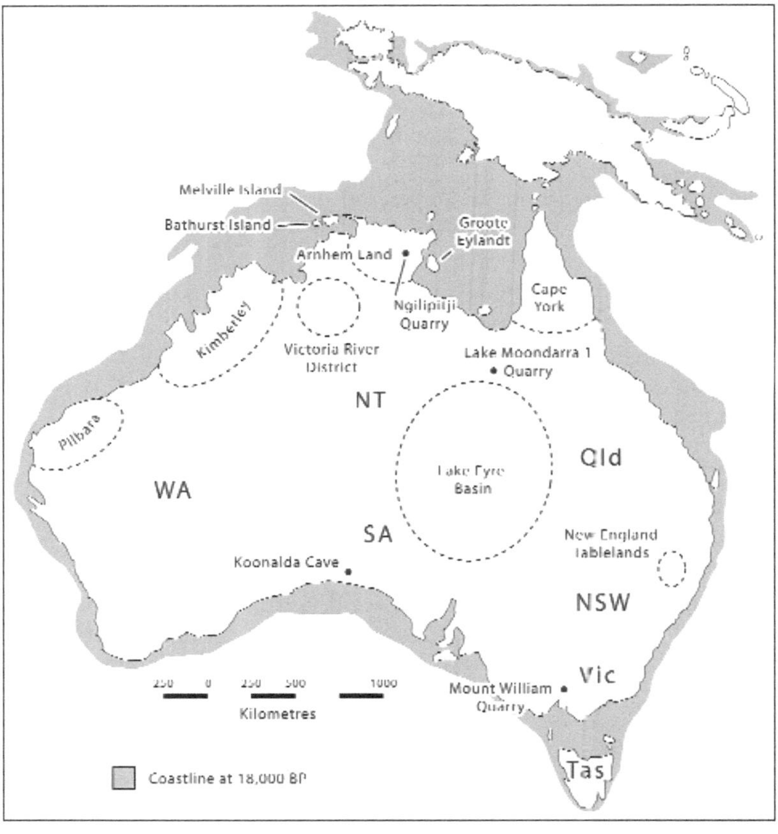

Figure 8.2: Map of Australia showing places mentioned in text.

than rigid dualism between the material and social worlds (Jordan 2001 and other chapters in Price 2001; see also Gell 1998; Kopytoff 1986).

In Australia, there is a strong correlation between the power of stone and the power of place (McBryde 1997; Taçon 1991). Where ethnography is available for ancient quarries, it is almost invariably the case that the mined rock outcrops are considered to be the creation of powerful Ancestral Beings (eg Flood 1997: 114; Graham and Thorley 1996: 84; Howitt 1904: 784; Jones 1990; Jones and White 1988; Palmer 1977; Paton 1994; Taçon 1991; Tindale 1965, 1974). In many cases, raw stones and minerals are thought to be transformed matter from the Dreaming, often the primary physical substances or emissions (ie blood, bones, fat, pus, feathers, and internal organs) of Ancestral Beings deposited onto the landscape and metamorphosed into rock. In eastern Arnhem Land, stone outcrops utilised as quarries are considered to be the 'flesh' of Ancestral Beings who are connected to certain moieties (D Leo, pers comm 2002). Coming from the bodies of mythical Dreaming figures, these stones and the artefacts made from them are believed to be charged with often dangerous Ancestral potency. In some areas, these are abstract religious concepts given corporeal representation (see Rappaport 1999: 141) through the phenomenology of certain natural properties of stone like colour and 'brilliance', which are understood to constitute an expression of Ancestral power (eg Palmer 1977; Taçon 1991). As Akerman (1995: 48) writes, stone objects are powerful even without significant technological modification because they have 'in effect, been consecrated from the time of their creation'.

STONE AXES IN ROCK ART AND MYTH

In caves and wet season shelters throughout the rock escarpments of western Arnhem Land (**Figure 8.2**), rock art researchers have recorded painted depictions of an anthropomorphic being with stone axes attached to or protruding from various parts of its body, primarily the ankles, knees, elbows, shoulders and genitals (Brandl 1982; Chaloupka 1993; Mountford 1955; see **Figure 8.3 colour section**). Such images are said by Aboriginal elders to depict Namarrgon the Lightning Man, or rather his transformed essence, spirit or 'shade' in the rock surface. Namarrgon is a dangerous Ancestral Being who uses stone axes to generate the lightning and thunder of wet season storms. He is said to travel through the dark monsoonal clouds, furiously smashing his stone axes together to produce lightning, or hurling them to earth, splitting trees and rocks, and frightening and sometimes killing humans (Chaloupka 1993: 58).

Reputedly, Arnhem Landers would even bury their own stone axes in the ground during electrical storms to avoid attracting the Lightning Man's fury (*ibid*: 38). (Interestingly, a similar custom was documented in the Brisbane region of south-east Queensland in the late 1800s amongst Aboriginal people, who hid their 'tomahawks' during lightning storms (Roth 1903: 8).) In contemporary Gunwinggu belief, Namarrgon's power is also said to come from his penis (Nganjmirra 1997), which is often prominently represented as a huge and grotesque phallus in rock paintings of the Ancestral Being (see Chaloupka 1993: 57, 58). Brandl notes one rock art motif in which the Lightning Man's penis is depicted as a stone axe, suggesting the male-oriented symbolic power and association of penises, stone axes and lightning (Brandl 1982: 63; see Sharp 1952 for details on the symbolic interrelationship between stone axes and male power in an Aboriginal society).

Most Complete Figure tradition (also known as 'X-ray') images of Namarrgon the Lightning Man belong to the recent Freshwater period of the Arnhem Land rock art sequence, and are thought to have been produced within the last 500 to 1,000 years (Taçon 1989). The most recent of these paintings was made by the last traditional Arnhem Land rock art painter, Najombolmi or 'Barramundi' Charlie (see **Figure 8.3 colour section**), in the late 1960s (*ibid*). Earlier images of less certain antiquity portray the Lightning Man as a therianthropic being and even as an hermaphrodite, depictions for which contemporary Aboriginal myths and stories apparently do not have an explanation (Chaloupka 1993). Chippindale *et al* (2000: 78) also note an archaic Dynamic tradition (perhaps 10,000-year-old) painting of a male figure with what appears to be a stone axe attached to or lying next to the ankle, raising the possibility that elements of the Lightning Man complex may have much older origins.

Rock paintings of stone axes are also associated with important mythic traditions, Ancestral Beings and lightning symbolism in other parts of northern Australia (Arndt 1962). In Bulajang or 'Sickness Country' in southern Arnhem Land, the Ancestral Being, Bula, is said by Jawoyn people to control the lightning as well as other powerful forces like bushfires, earthquakes and volcanoes (Gunn 1992: 174). Bula is often depicted in the rock art of this region as an anthropomorphic figure holding a stone axe (Welch 1992: 196). Jawoyn elders describe Bula as an extremely dangerous Ancestral Being who if suitably disturbed has the power to destroy the world (Gunn 1992: 176).

The well known Lightning Brothers site on Delamere Station in the Victoria River District (VRD) (**Figure 8.2**), or the 'Land of the Lightning Brothers', features two spectacular painted and engraved striped anthropomorphic figures, one with a stylised depiction of a stone axe held under its arm (Arndt 1962; Walsh 1988: 200). One of the many Lightning Brothers myths, as recounted by Wardaman people, explains this implement to be the powerful lightning-generating stone axe used by the furious elder brother, Yagjagbula, to decapitate his wife-stealing younger brother Yabiringi (Walsh 1988: 76). Another famous rock art site in the VRD related to the Lightning Brothers mythological complex is located at Ingilladdi, and features painted striped anthropomorphic figures holding stone axes (*ibid*: 76). At least seven other art sites with painted Lightning Brothers figures are known in the VRD (McNickle 1991: 39). In addition to the Lightning Brothers complex, 10 other sites in the VRD have been documented that feature paintings of large, horned anthropomorphic beings with stone axes extending from their fingers. These regionalised 'stone axe men' motifs have apparent mythological overtones, and are described by McNickle (1991: 40) as 'among the most visually impressive of VRD motifs'.

There are further parallels between Ancestral Beings, stone axes and lightning in the mythologies of the wider northern Australian region. The Thundering Gecko is an important Ancestral Being in the oral traditions of the Nunggubuyu people of south-eastern Arnhem Land (Van der Leeden 1975: 85). In one myth, the Thundering Gecko is busy sharpening his stone axe when he is heard by a malevolent Spirit. Drawn by the sound of axe-grinding, the Spirit ominously approaches but is spied by the Thundering Gecko, who uses his stone axe to flash lightning at the Spirit. After a running battle during which the Spirit is repeatedly struck by lightning whilst trying to spear the Thundering Gecko, the heroic Ancestral Being buries his stone axe in the ground and disappears into a waterhole at Anarbarin, an important Dreaming site for the Nunggubuyu (*ibid*: 85).

Amongst the Tiwi people of Bathurst and Melville Islands (**Figure 8.2**), it is a feared female Ancestral Being, Bumerali the Lightning Woman, who controls the lightning and thunder of the monsoonal wet season (Mountford 1958: 44–45; 1973: 66). Bumerali is believed by Tiwi people to travel through the thunderclouds with her husband and many children. With stone axes hafted onto 'long crooked handles' (perhaps symbolic of lightning bolts) she strikes the ground beneath her, splitting trees and killing humans (*ibid*: 44). The sharp clap of thunder is thought to be the sound of Bumerali angrily rebuking her children as they disobediently jump from cloud to cloud. Along with her brother Pakadringa, 'the man of the thunderstorms', and her sister Tomituka, 'the woman of the monsoon rain', this family of Ancestral Beings brings the life-renewing rains of the wet season to the Tiwi people (*ibid*).

Associations between stone axes, Ancestral Beings and lightning themes in Aboriginal cosmologies are relatively common only in the Northern Territory region of northern Australia (Arndt 1962). But despite this restricted distribution, they appear to have been spreading elsewhere in recent times. Kim Akerman (pers comm 2000) notes the contemporary use of steel hatchets symbolising the lightning-generating stone axes of Ancestral Beings in parts of the great Ngawayil rain-making ceremonies of the eastern Pilbara and adjacent desert regions of Western Australia. In these ritual performance episodes, steel hatchets are clashed together by dancing actors re-enacting a fight between furious storm beings, with the resulting sparks symbolic of lightning bolts. As Akerman suggests, this ceremony appears to have moved into the region over the last 100 years or so, and may be a reason for the few exotic forms of stone axes that appear sporadically over otherwise non-axe areas of Western Australia.

STONE AXES IN RITUAL

Due to their replacement with metal tools in even the most remote regions of Aboriginal Australia by the 1930s or so, first-hand accounts of the use of stone axes are unfortunately rare (Dickson 1981). However, fragmentary documentary evidence for the ritualistic use of stone axes does provide important insights into their former social and symbolic roles (see especially Sharp 1952; Stanner 1960:

121). The use of stone axes in ritual and symbolic contexts appears to have been widespread in Australia in the recent ethnographic past.

Stone axes were especially prominent in burial ceremonies. For example, in 1911, Baldwin Spencer documented Kakadu (Gunwinggu) people of central Arnhem Land displaying stone axes during a ritual mourning ceremony, despite the fact that, as he observed, 'nowadays … the old stone axe is practically a thing of the past' (Spencer 1966: 242). In the late 1800s, Spencer (1928: 489–91) also made a photographic record of a Warramungu burial rite in central Northern Territory.[1] During this ceremony, a paperbark bundle containing an arm bone of the deceased was broken by a male elder with a large hafted stone axe, a dramatic event which caused female mourners to scatter screaming from the burial site. Male and female mourners were also known to gash open their scalps in sorrow with stone axes during burial ceremonies (eg Hart *et al* 1988: 96; Roth 1907: 396). In another death-related ceremony, Aboriginal women in the eastern coastal region of central Queensland used stone axes to split the thighbones of deceased kin in order to determine who had caused their death by sorcery, with the sound of splintering bone implicating the name of the murderer (Roth 1907: 400).

Stone axes also appeared in other ritual contexts. For example, 'clever' men (for discussion, see below) of the Moreton Bay region in south-east Queensland were reputed to obtain ritual paraphernalia such as quartz crystals and also bring rain by diving into deep waterholes, where the rainbow was said to dwell, and cutting it free with a 'tomahawk' (Petrie 1904: 186). Stone axes were also used in this region to break the earth during the construction of sacred ceremonial grounds or Bora rings, and to cut open the first bunya pine nuts of the season in a ritual invoked to ensure the continued abundance of this valued resource (*ibid*: 15, 49). Axes were reputedly involved in another religious ceremony in the south-east Queensland area in which the tools were used to chip small pieces of stone from a large rock (whilst simultaneously conversing with it) believed to be the petrified body of a noted Ancestral Being (Steele 1983: 42). On Groote Eylandt, stone axes were used for cutting into the sacred Maitjunaua bloodwood tree to free the north-west wind in important wet season 'increase ceremonies' (Mountford 1956: 96; 1973: 172). Finally, Sharp (1952: 416) recorded stone axes being employed as important totemic symbols in ritual performance episodes of the Sunlit Iguana clan of the 'Yir Yoront' in western Cape York.

Analysis of the archaeological literature also reveals that axes were sometimes included as grave goods with human burials in a variety of regions in Australia, most dated to within the last few thousand years (eg Flood 1980: 215; Haglund 1976; Howitt 1904: 465; Massola 1968: 132–33). In south-east Australia, the interred axes appear in some cases to have been ritually broken or 'killed' prior to deposition with bodies (Massola 1968), although whether the destruction of axes took place as a direct part of funerary rites or otherwise is not known. Accordingly, even axe fragments may have had symbolic significance, perhaps because they retained some of the essence and power of the source of origin for the stone, and/or the tool's owner. For instance, axe flakes were apparently placed in the facial region of an individual buried at the large Broadbeach cemetery in south-east Queensland, and axe flakes (as well as an undamaged polished axe) were also found with other interments at the burial site (Haglund 1976: 28, 75, 76). These burials are around 1,000 years old. Roth (1903: 33) recorded 'clever' healers on the Bloomfield River in Cape York removing stone axe flakes from the bodies of sick people. These flakes were dangerous objects believed to have been magically inserted into the victim's body by malevolent sorcerers. Presumably, then, they were also the ritual paraphernalia of 'clever' people secreted on their person and produced during healing ceremonies.

Ethnographic evidence also points to the curation of axe flakes for apparently non-functional purposes. For instance, in Victoria in 1862, probably half a century after operations at the massive Mount William greenstone axe quarry (see McBryde 1978; **Figure 8.2**) had ceased, the Victorian pioneer Isaac Batey (Batey 1885: 157–58) encountered a Wiradjuri man carrying around a 'fragment of a broken stone hatchet' he had recently procured from the extensive site. Significantly, it would appear that Aboriginal people were making visits to this quarry to collect stone many years after it had been abandoned, and despite the functional obsolescence of edge-ground axes. A story from northern Arnhem Land concerning the creation of the first sacred 'bull-roarer' even mentions that the ritual musical object (held to be symbolic of Ancestral Beings (Elkin 1938: 208) and used in ceremony to provide a direct link to the Dreaming) was made from a 'chip' from a stone axe (Marshall 1952: 101–04). Ideas of the symbolic significance of axe fragments may explain why some archaeological specimens were effectively destroyed by extensive reworking episodes apparently without good technological or functional cause (M Moore, pers comm 2002).

'CLEVER' PEOPLE AND STONE AXES

In Aboriginal Australia, religious specialists recognised by their communities as having specially developed relationships with the supernatural world are referred to as 'clever' people (Elkin 1994; Hume 2002). In Arnhem Land, such shamanic figures are said to be able to call upon the power of Namarrgon the Lightning Man through sorcery (Isaacs 1980: 129). For example, an oral narrative from Jawoyn country refers to an old 'clever' man constructing a human-like figure of Marrgon the Lightning spirit from grass and fibre-string, and attaching stone axes to its knees, elbows and shoulders (Robinson 1966: 140–43). He then 'sings' (infuses with ritual power through mystical incantation) the Lightning spirit to life and sends it after his wife, who has run away with a younger man. In a dream, the 'clever' man sees the Lightning spirit stalk and kill the fleeing couple, whose mutilated remains he later finds. A description of the Lightning spirit was offered by the Aboriginal narrator:

> No matter if it is a long way off that anyone sits down, that lightning will find them ... He has stone-axe along shoulder, knee, elbow. 'Rai, rai, dee-dee, Siss-siss,' he talks as he strikes, as he cuts anyone up. (*ibid*: 143)

In a similar story recounted by Gunwinggu people of western Arnhem Land, an old 'clever' man is taken into the sky by spirits and taught ritual knowledge by Namarrkon the Lightning Man, after which he produces a rock art depiction of the being during a special ceremony (Godden and Malnic 1997: 50–51). When the 'clever' man's wife absconds with another man, he 'sings' them both at this rock art site by constructing a human-like figure from grass and attaching miniature stone axes to it. This ritual summons the enraged Namarrkon from the sky, who upon tracking down the couple, snatches a stone axe from one of his knees and blasts them both to death with sheets of fire.

Stone axes were associated with 'clever' practices and knowledge elsewhere in northern Australia. Warner (1969: 188–90) makes note of an interview with a Yolngu 'clever' man from eastern Arnhem Land in the 1930s. According to the testimony of this informant, during a ritual killing, he had used his axe/hatchet to strike a female victim dead between the eyes. He then removed the handle from the tool and used it to create a hole in her body through which he placed his hand and removed her heart. After he had magically sealed the wound, the 'clever' man whirled the bloodied handle around her head, bringing the woman back to life only for her to die suddenly some time later. Warner (*ibid*: 205) also notes the case of another Yolngu man who was hunting in the bush when he was pulled into a water-hole by spirit familiars, who made him 'clever'. After this supernatural transformation the man struck each of the familiars on the nose with his stone axe and then healed them with his new powers.

Finally, Taçon (1991: 199) mentions a stone axe quarry in north-western Arnhem Land known to Aboriginal elders as Thunder Rock. This dangerous place is said to attract lightning. A person's death could be caused there by sorcery by cutting a vine that grew in the quarry vicinity and calling out the victim's name.

AXE EXCHANGE AND THE ORIGINS OF THE DREAMING

Australian exchange systems of the mid to late Holocene period were amongst the most extensive and complex found anywhere in the world (McBryde 1987, 1997). In the recent ethnographic past, regional exchange networks saw the distribution of valued material goods like stone, ochre, pearl shell and the psychotropic drug *pituri* throughout most of Aboriginal Australia (Davidson *et al* forthcoming; McBryde 1987, 1997; McCarthy 1977; Roth 1897; Stanner 1933; Thomson 1949). To a certain extent, social benefits extending from ritual gift exchange systems seem to have outweighed mundane economic factors, particularly with regards to trade in stone tools (see especially Paton 1994). As the anthropologist Stanner (1933: 170) commented, '[t]here can be nothing more useless in the native economy than a pile of foreign spears or axe-heads ... with no where to send them'. Furthermore, archaeological evidence implies that, in the Lake Eyre Basin region of central Australia and in central Victoria, south-east Australia, axes from certain major quarries were traded into areas where local production of axes took place, or where there was no shortage of suitable raw materials (Davidson *et al* forthcoming; cf McBryde 1978 for south-east Australia). This would seem to indicate that prehistoric axe exchange also involved more than mundane utilitarian considerations.

Recent research suggests that stone axes from some major quarries were traded up to 800 to 1,000 km from their source of origin in central and southern Australia (McBryde 1987). The largest of these quarries, for example Lake Moondarra 1 in north-west central Queensland (see **Figure 8.2**) – which supplied axes into the Lake Eyre Basin region (Roth 1897; Tibbett 2002) – are referred to by Davidson *et al* (forthcoming) as Long Transfer Quarries. At present indication, the complex inland exchange systems associated with these intensively worked quarries appeared roughly 1,000 years ago (Hiscock 1988).

Although stone for making tools is not widely traded today, Aboriginal people involved in historically recorded exchange systems were generally aware of the precise location on the distant landscape where exotic stone tools came from (eg Tindale 1974: 86). Clearly, traded stone tools were not unknown alien objects, but things with known historical meanings and a place of origin within Aboriginal cosmological worldviews. Even at distances of up to several hundreds of kilometres away from the quarry, Aboriginal people were in some cases aware of the ritual knowledge for stone sources and associated mythic traditions, songs and ceremonies (see McBryde 1987, 1997; Tindale 1974: 86). Such an intimate level of petrological knowledge suggests the important role that the exchange of stone artefacts as 'pieces of places' (see Bradley 2000: 88) has played in creating and mediating symbolic human-landscape relations in Australia. Consider Gould's (1977: 164) observation of stone use in the central desert region of Western Australia:

> ... some quarries occur at or near sacred sites – that is, totemic 'dreaming' places. People who believe themselves to be descended patrilineally from the particular totemic being at one of these sites will make special trips to the quarry to secure the stone there. A man places a high value on stone from a site of his dreamtime totem. Stone like this is often transported over long distances (as much as 500 km) and is given to distant kinsmen of the same totemic patrilineage ... Because of his patrilineal relationship to the site, a man sees the stone as part of his own being ...

In this arid region, stone from a known totemic source was traded to distant kin as a symbol of interrelatedness between people and place (Gould 1977). Aboriginal men regarded raw stone materials from particular quarries as a natural part of their own being and Dreaming identity. Similar analogies may be drawn with other areas of Australia where stone was widely traded (Jones 1990; Jones and White 1988; Paton 1994). But is it possible to place a time-depth on such place-oriented lithic symbolism, particularly that involving axes?

Social mechanisms for the procurement of axe stone may have existed during the Pleistocene period. Schrire (1982: 241) excavated a cache of several complete stone axes from the lowest level of Nawamoyn rockshelter in northern Arnhem Land. These unused artefacts appear to have been hidden in a recess at the back of the shelter around 20,000 years ago, perhaps implying deliberate storage for later exchange (see Hiscock 1988). Moreover, Jones and Johnson (1985: 217) argue that 25,000–30,000 year-old stone axes excavated from sites in western Arnhem Land were manufactured from porphyritic dolerite procured from sources a 'minimum distance of several scores of kilometres' from the study area, suggesting movement of axe stone over considerable physical distances.

Ancient rock art may provide further important, if elusive, clues. The oldest known depictions of stone artefact use in Australian rock art are hafted axes in the Dynamic Figure tradition of western Arnhem Land, thought to be at least 10,000 years old (Taçon and Brockwell 1995: 687–89; see **Figure 8.4 colour section**). Some Dynamic paintings portray stone axes in scenes reminiscent of those argued by Chippindale *et al* (2000) to contain a strong visionary/shamanistic element. For example, one composition from Deaf Adder Creek features a human male (associated with a small therianthropic figure) armed with an upraised stone axe as though about to strike a large snake-like being surrounded by wild bees (Flood 1995: 169). If Chippindale *et al* (2000) are correct in their shamanistic interpretation of Dynamic art, then these very old and evocative images offer tantalising glimpses into the possible antiquity of stone axe symbolism in northern Australia, in particular the tool's contemporary links with mythical Ancestral Beings and the practices and knowledge of 'clever' people or shamans.

Despite the potential symbolic content of these very ancient images, paintings of anthropomorphic beings with stone axes protruding from their bodies – interpreted by Arnhem Land elders to be images of Namarrgon the Lightning Man – are probably only around 500 to 1,000 years old (P Taçon, pers comm 2002). Furthermore, archaeological research in the VRD indicates that painted anthropomorphic figures interpreted by Wardaman elders to be images of the Lightning Brothers may also date to around 1000 BP (Flood 1997: 307). Interestingly in this regard, Ross (1997; see also Davidson *et al* forthcoming) argues that a regionally distinctive body of anthropomorphic/therianthropic rock art motifs in the Mt Isa region of north-west central Queensland is associated with the emergence of major Long Transfer axe quarries and exchange systems in the Lake Eyre Basin area roughly 1,000 years ago.

The appearance of these distinctive art traditions in several regions of late Holocene northern Australia suggests that significant changes in the symbolic meanings of stone and stone axes may have occurred in relatively recent times. These changes seem to have revolved around the emerging symbolic association of stone axes with images of supernatural/mythological figures interpreted by modern Aboriginal elders to be powerful Ancestral Beings. Ancestral Beings are the First People of the Aboriginal Dreaming, creationary heroes whose epic journeys across the once shapeless world resulted in the formation of the present-day landscape and the creation of all human, animal and plant life within it (Morphy 1989). Although more detailed research would be required to test such an inference, the merging of stone axe imagery with Ancestral Being themes in recent rock art traditions could allude to the possibility that cultural transformations in the stone axe's meanings may be explicitly linked to the historical origins of the Aboriginal Dreaming in Australia within the last few thousand years. Indeed, as David (2002) persuasively argues, the Dreaming as it is known in the ethnographic period may have an antiquity of less than 1,400 years.

Certainly, in the recent ethnographic past, the mythological pathways Ancestral Beings travelled between various sources of stone and other sacred Dreaming sites, known as Dreaming Tracks, were interchangeable with the historical trading routes along which stone was traditionally exchanged between different territorial estates or 'countries' (McBryde 1997; McCarthy 1977). In the past, quarries whose axes were distributed in specific directions for hundreds of kilometres across the landscape – such as the Selwyn Ranges' metabasalt quarries in the north-west central Queensland highlands and those of the New England Tablelands, the extensive quarries of north-west New South Wales, and the intensively quarried outcrops in the Cambrian greenstone belt of central Victoria – may have been places of particularly heightened symbolic power in the emerging religious system of the Dreaming. Accordingly, the initial movement of stone axes (and other lithic artefacts, for example microlithic tools, stone knives and projectile points) into areas where they were previously unknown or rare may even have contributed to the historical *creation* of many of the major mythological Dreaming Tracks across the Australian landscape (see especially McCarthy 1977: 252; cf Evans and Jones 1997). Given the appearance of the most large-scale of Australia's Long Transfer quarries within the extensive exchange system of the Lake Eyre Basin region around 1,000 years ago (Davidson *et al* forthcoming; Tibbett 2002), and the correspondingly intensive movement of stones, stories and symbols for many hundreds of kilometres across the Australian landscape, this is a dynamic process which arguably may have intensified dramatically in the last millennium or so.

Even if the Dreaming as it is known today may be less than 1,400 years old (David 2002), it seems likely that such a complex belief system emerged from existing localised worldviews in which stone and stone artefacts comprised more than mundane matter and materiality for Aboriginal people (see, for example, Fullagar and Head 1999). The much overlooked stone arrangements, standing stones and rock art at several major quarries in Australia (Knuckey and Kippen 1992; McBryde 1973, 1997; Mulvaney 2001; O'Connell 2000; Towle 1939), whilst of uncertain antiquity, highlight the possibility that on-source stone extraction has long been imbued with symbolic values.

SUMMARY

In this chapter, I have employed as a case example the Aboriginal edge-ground stone axe, a near ubiquitous hafted tool type with a Pleistocene antiquity in tropical northern Australia, to demonstrate the salient socio-symbolic roles of supposedly 'ordinary' functional stone artefacts in past Aboriginal societies and exchange systems (cf Smith 1999). Via closer consideration of rock art, ethnography, oral tradition and archaeological sources, it has been argued that stone axes are not simply straightforward utilitarian objects for Aboriginal people, past and present. Despite fulfilling 'ordinary' functional roles, stone axes are equally important ritual 'power tools' associated with a wide range of socially significant practices, including sorcery, shamanism, and burial rites. In several areas of Australia, axes are also explicitly associated by Aboriginal people with the dangerous power of Ancestral Beings and other supernatural and mythological figures, some of whom use axes to generate lightning and thunder. Ancient rock art evidence from various parts of northern Australia would indicate that these symbolic associations have some antiquity, at least within the late Holocene period but possibly extending earlier into the prehistory of the continent as well.

In recent Aboriginal societies, the intimate petrological knowledge of individuals within long-distance exchange networks shaped human social and symbolic interactions with raw stone materials, and informed the particular nature of the exchange of lithic artefacts between people. This complex system of stone symbolism appears to have been conceptualised and mediated with reference to the permeating reality of the Dreaming, a potentially historically recent (in terms of its ethnographic form) landscape-focused cosmology (David 2002) that attributes certain key stone sources and other sites in the landscape with Ancestral power (see also Taçon 1991). Recent changes in rock art depictions of stone axes and the emergence of intensive long-distance stone axe exchange networks (including the appearance of Long Transfer quarries) in the mid to late Holocene period, but especially within the last 1,000 years or so, may be correlated (see Ross 1997), and perhaps even directly linked to the historical origins of the modern Aboriginal Dreaming.

CONCLUSIONS

On a final note, readers may have noticed that the Aboriginal edge-ground axe has been referred to in this chapter as though there was only one category of the tool in ancient Australia: *the* stone axe. This is problematic, because even a brief examination of archaeological specimens from around the continent reveals the wide range of technological and morphological variation in axe forms (Dickson 1981; McCarthy 1976). In recent times some Aboriginal people also had different names for small and large stone axes (eg Evans and Jones 1997: 401), suggesting that various 'styles' of axe may have carried alternative meanings. Limitations of space unfortunately do not permit a more detailed treatment of this issue, but an interesting subject for further research may involve an analysis of the potentially differing symbolic values, social uses, cultural biographies (see Kopytoff 1986) and depositional histories that various types or shapes of stone axe may have had in ancient Australia, with a particular emphasis on size, elaborateness of form, and technological variation through time and space, perhaps on an intra- and inter-regional scale (see Akerman *et al* 2002). A consideration of rock art depictions of stone axes, in particular negative stencils of the tools, may provide an interesting depth to this study.

Of even greater interest, however, in light of the theme of this volume, might be to explore the notion that particular meanings imposed through the human modification of raw materials – or rather, the act of releasing the axe from within the unshaped block of stone, as knapping is sometimes perceived in Aboriginal society (eg Graham and Thorley 1996; Jones and White 1988; Paton 1994) – may have been transcended to a certain degree by the underlying symbolic values or power of the raw stone materials from which axes and other artefacts were made (see Akerman 1995: 18). Of course, the particular shape of a stone axe and the specific way it was worked can be expected to have conveyed multiple social meanings for Aboriginal people, especially during the performance of manufacture (see Dobres 2000). But the rich Australian ethnography also tells us that in terms of wider philosophical understandings of the Dreaming landscape and the Ancestral forces responsible for its creation, it may be the raw, immutable substance or mineralogical matter from which stone objects are made that holds the deepest symbolic values. These are cultural notions bound through real and mythical time to the profundity of the Dreaming and the sacredness of landscape in everyday Aboriginal understanding, to the power that lingers naturally in stone due to the creative actions of Ancestral Beings. In many ways, the antiquity of stone axe symbolism in ancient Australia, evidenced by rock art and the intensive exchange of axes made from a range of distinctive raw material types extracted from particular places within the landscape, probably serves to demonstrate as much the meaningfulness of stone itself as the axes it was used to make in Aboriginal prehistory.

Accordingly, if we consider, together with the archaeological data, the ethnographic evidence for the place-oriented nature of recent Aboriginal interactions with rocks and minerals – whereby some stones are understood through various aspects of their materiality to embody the symbolic potency of certain powerful places within the landscape (see also Bradley 2000; Cooney 1998; Edmonds 1995), regardless of what humans do to them technologically – then it becomes clear that it is patterns of *raw material* usage through time and space that may provide some of the most compelling insights into ancient social and cognitive systems, cosmological beliefs and ideologies in Australia, and historical changes within these entities.

Note

1 This photograph cannot be reproduced due to the culturally sensitive nature of the imagery.

Acknowledgments

First and foremost I would like to thank Paul Taçon for his guidance, insight and enlightening conversations in the field, and for suggesting the title to this paper. John Campbell, Ann Wallin, Michael Strong, Douglas Hobbs, Mark Moore,

Michael Morrison, Kim Akerman, Paul Memmott, Paul Crook, Daniel Leo, Graham Knuckey and Fiona Leslie are also acknowledged for their assistance and advice. Nicky Boivin and Mary Ann Owoc are kindly thanked for their invitation to participate in the present volume, and for making various helpful comments on earlier drafts that much improved my writing. 'Muscly' Tommy, Ruby, Colin and Ian Saltmere, Bruno Martin, Danny Butcher, Georgie Anderson and other Indjilandji, Alyawarr, Waanyi, Warlpiri and Warramungu people of north-west Queensland and beyond provided the central source of inspiration for this work, and I thank them all for their teachings, kindness and hospitality. Despite the input of all these people, however, only I am to blame for any errors or shortcomings in this paper, and my views do not necessarily represent theirs.

References

Akerman, K (1995) 'Tradition and change in aspects of contemporary Australian Aboriginal religious objects', in Anderson, C (ed), *Politics of the Secret*, Oceania Monographs 45, Sydney: University of Sydney

Akerman, K, Fullagar, R and van Gijn, A (2002) 'Weapons and *Wunan*: production, function and exchange of Kimberley points' 1 *Australian Aboriginal Studies* 14–42

Arndt, W (1962) 'The interpretation of the Delamere lightning painting and rock engravings' 32(3) *Oceania* 163–77

Batey, I (1885) 'Notes on the habits of native birds' 2(7) *The Victorian Naturalist* 157–58

Bernstein, JH (1997) *Spirits Captured in Stone: Shamanism and Traditional Medicine Among the Taman of Borneo*, London: Lynne Rienner

Bradley, R (2000) *An Archaeology of Natural Places*, London: Routledge

Brandl, EJ (1982) *Australian Aboriginal Paintings in Western and Central Arnhem Land: Temporal Sequences and Elements of Style in Cadell River and Deaf Adder Creek Art*, Canberra: Australian Institute of Aboriginal Studies

Chaloupka, G (1993) *Journey in Time: The World's Longest Continuing Art Tradition, the 50,000-Year Story of the Australian Aboriginal Rock Art of Arnhem Land*, Chatswood: Reed

Chippindale, C, Smith, B and Taçon, PSC (2000) 'Visions of dynamic power: archaic rock-paintings, altered states of consciousness and "clever men" in western Arnhem Land (NT), Australia' 10(1) *Cambridge Archaeological Journal* 62–101

Cooney, G (1998) 'Breaking stones, making places: the social landscape of axe production sites', in Gibson, A and Simpson, D (eds), *Prehistoric Ritual and Religion*, Phoenix Mill: Sutton

David, B (2002) *Landscapes, Rock-Art and the Dreaming: An Archaeology of Pre-Understanding*, Leicester: Leicester University Press

Davidson, I (1999) 'Symbols by nature: animal frequencies in the Upper Palaeolithic of Western Europe and the nature of symbolic representation' 34 *Archaeology in Oceania* 121–31

Davidson, I, Cook, N, Fischer, M, Ridges, M, Ross, J and Sutton, S (forthcoming) 'Archaeology in another country: exchange and symbols in north west Central Queensland', in Macfarlane, I, Paton, R and Mountain, M-J (eds), *Many Exchanges: Archaeology, History, Community and the Work of Isabel McBryde*, Canberra: Aboriginal History Inc

Dickson, FP (1981) *Australian Stone Hatchets: A Study in Design and Dynamics*, Sydney: Academic

Dobres, MA (2000) *Technology and Social Agency*, London: Blackwell

Edmonds, M (1995) *Stone Tools in Society: Working Stone in Neolithic and Bronze Age Britain*, London: BT Batsford

Elkin, AP (1938) [1981] *The Australian Aborigines*, London: Angus and Robertson

Elkin, AP (1994) *Aboriginal Men of High Degree: Initiation and Sorcery in the World's Oldest Tradition*, St Lucia: University of Southern Queensland

Evans, N and Jones, R (1997) 'The cradle of the Pama-Nyungans: archaeological and linguistic speculations', in McConvell, P and Evans, N (eds), *Archaeology and Linguistics: Aboriginal Australia in Global Perspective*, Oxford: OUP

Flood, J (1980) *The Moth Hunters: Aboriginal Prehistory of the Australian Alps*, Canberra: Australian Institute of Aboriginal Studies

Flood, J (1995) *Archaeology of the Dreamtime: The Story of Prehistoric Australia and its People*, Sydney: Angus and Robertson (revised edn)

Flood, J (1997) *Rock Art of the Dreamtime: Images of Ancient Australia*, Sydney: Angus and Robertson

Fullagar, R and Head, L (1999) 'Exploring the prehistory of hunter-gatherer attachments to place: an example from the Keep River Area, Northern Territory, Australia', in Ucko, PJ and Layton, R (eds), *The Archaeology and Anthropology of Landscape*, London: Routledge

Gell, A (1998) *Art and Agency: An Anthropological Theory*, Oxford: Clarendon

Gero, JM (1989) 'Assessing social information in material objects: how well do lithics measure up?', in Torrence, R (ed), *Time, Energy and Stone Tools*, Cambridge: CUP

Godden, E and Malnic, J (1997) *Rock Paintings of Aboriginal Australia*, Victoria: Reed

Goldsworthy, A (1994) *Stone*, Harmondsworth: Penguin

Gould, RA (1977) 'Ethno-archaeology: or, where do models come from?', in Wright, RVS (ed), *Stone Tools as Cultural Markers: Change, Evolution and Complexity*, New Jersey: Humanities Press

Graham, R and Thorley, P (1996) 'Central Australian Aboriginal stone knives: their cultural significance, manufacture and trade', in Morton, SR and Mulvaney, DJ (eds), *Exploring Central Australia: Society, the Environment and the 1894 Horn Expedition*, Chipping Norton: Surrey Beatty and Sons

Graves-Brown, PM (1995) 'Fearful symmetry' 27(1) *World Archaeology* 88–99

Graves-Brown, PM (2000) *Matter, Materiality and Modern Culture*, London: Routledge

Gunn, RG (1992) 'Bulajang – a reappraisal of the archaeology of an Aboriginal religious cult', in McDonald, J and Haskovec, IP (eds), *State of the Art: Regional Rock Art Studies in Australia and Melanesia*, Melbourne: Australian Rock Art Research Association

Haglund, L (1976) *The Broadbeach Aboriginal Burial Ground: An Archaeological Analysis*, St Lucia: University of Queensland

Hampton, OW (1999) *Culture of Stone: Sacred and Profane Uses of Stone among the Dani*, College Station: Texas A and M University Press

Hart, CWM, Pilling, AR and Goodale, JC (eds) (1988) *The Tiwi of North Australia*, New York: Holt, Rinehart and Winston

Harvey, G (2000) 'Art works in Aotearoa', in *Indigenous Religions: A Companion*, London: Cassell

Hayden, B (1979) *Palaeolithic Reflections: Lithic Technology and Ethnographic Excavation among Australian Aborigines*, Canberra: Australian Institute of Aboriginal Studies

Hiscock, P (1988) 'A cache of tulas from the Boulia district, Western Queensland' 23 *Archaeology in Oceania* 60–70

Howitt, AW (1904) [1996] *The Native Tribes of South-East Australia*, London: Macmillan (republished 1996, Canberra: Aboriginal Studies Press)

Hume, L (2002) *Ancestral Power: The Dreaming, Consciousness and Aboriginal Australians*, Carlton South: Melbourne University Press

Imamura, K (1996) *Prehistoric Japan: Perspectives on Insular East Asia*, London: UCL Press

Isaacs, J (1980) *Australian Dreaming: 40,000 Years of Aboriginal History*, Sydney: Lansdowne

Jones, R (1990) 'Hunters of the Dreaming: some ideational, economic and ecological parameters of the Australian productive system', in Yen, D and Mummery, J (eds), *Pacific Production Systems: Approaches to Economic Prehistory*, Canberra: Australian National University

Jones, R and Johnson, I (1985) 'Deaf Adder Gorge: Lindner Site, Nauwalabila 1', in Jones, R (ed), *Archaeological Research in Kakadu National Park*, Canberra: Australian National Parks and Wildlife Service

Jones, R and White, N (1988) 'Point blank: stone tool manufacture at the Ngilipitji Quarry, Arnhem Land, 1981', in Meehan, B and Jones, R (eds), *Archaeology with Ethnography: An Australian Perspective*, Canberra: Australian National University

Jordan, P (2001) 'The materiality of shamanism as a "world-view": praxis, artefacts and landscape', in Price, NS (ed), *The Archaeology of Shamanism*, London: Routledge

Kenoyer, JM (1993) 'Socio-ritual artifacts of upper Palaeolithic hunter-gatherers in South Asia', in Possehl, G (ed), *South Asian Archaeology Studies*, New York: International Science Publisher

Knuckey, G and Kippen, K (1992) 'The archaeology of Stonehenge – a preliminary survey' 9 *Queensland Archaeological Research* 17–25

Kopytoff, I (1986) 'The cultural biography of things: commoditization as process', in Appadurai, A (ed), *The Social Life of Things: Commodities in Social Perspective*, Cambridge: CUP

McBryde, I (1973) 'Stone arrangements and a quartzite quarry site at Brewarrina' 9 *Mankind* 118–21

McBryde, I (1978) '"Wil-im-ee Moor-ring" or, where do Axes come from?' 11 *Mankind* 354–82

McBryde, I (1984) 'Kulin greenstone quarries: the social contexts of production and distribution for the Mt William site' 16 *World Archaeology* 267–85

McBryde, I (1986) 'Artefacts, language and social interaction: a case study from south-eastern Australia', in Bailey, GN and Callow, P (eds), *Stone-Age Prehistory: Studies in Memory of Charles McBurney*, Cambridge: CUP

McBryde, I (1987) 'Goods from another country: exchange networks and the people of the Lake Eyre basin', in Mulvaney, DJ and White, JP (eds), *Australians to 1788*, Broadway: Fairfax, Syme and Weldon

McBryde, I (1997) 'The landscape is a series of stories. Grindstones, quarries and exchange in Aboriginal Australia: a Lake Eyre case study', in Ramos-Millan, A and Bustillo, MA (eds), *Siliceous Rocks and Culture*, Granada: University of Granada

McCarthy, FD (1976) *Australian Aboriginal Stone Implements: Including Bone, Shell and Tooth Implements*, Sydney: Australian Museum Trust

McCarthy, FD (1977) 'The use of stone tools to map patterns of diffusion', in Wright, RVS (ed), *Stone Tools as Cultural Markers: Change, Evolution and Complexity*, New Jersey: Humanities Press

McNickle, HP (1991) 'A survey of rock art in the Victoria River district, Northern Territory' 8(1) *Rock Art Research* 36–46

Marshall, A (1952) [1978] *People of the Dreamtime*, Melbourne: Hyland House (republished 1978)

Massola, A (1968) *Bunjil's Cave: Myths, Legends and Superstitions of the Aborigines of South-East Australia*, Melbourne: Lansdowne

Moore, MW (2003) 'Flexibility of stone tool manufacturing methods on the Georgina River, Camooweal, Queensland' 38(1) *Archaeology in Oceania* 23–36

Morphy, H (1989) 'On representing Ancestral Beings', in *Animals into Art*, London: Unwin Hyman

Morphy, H (1999) 'Encoding the Dreaming' 49 *Australian Archaeology* 13–22

Morwood, MJ (2002) *Visions from the Past: The Archaeology of Australian Aboriginal Art*, Crow's Nest: Allen and Unwin

Morwood, MJ and Hobbs, DR (1995) 'Themes in the prehistory of tropical Australia', in Allen, J and O'Connell, JF (eds), *Pleistocene to Holocene in Australia and Papua New Guinea, Transitions: Antiquity* special no 265, 747–68

Morwood, MJ and Trezise, PJ (1989) 'Edge-ground axes in Pleistocene Greater Australia: new evidence from SE Cape York Peninsula' 6 *Queensland Archaeological Research* 77–90

Mountford, CP (1955) 'The Lightning Man in Australian mythology' 55 *Man* 129–30

Mountford, CP (1956) *Records of the American-Australian Scientific Expedition to Arnhem Land: Art, Myth and Symbolism*, Melbourne: Melbourne University Press

Mountford, CP (1958) *The Tiwi: Their Art, Myth and Ceremony*, London: Phoenix House

Mountford, CP (1973) *The Dreamtime Book: Australian Aboriginal Myths*, Adelaide: Rigby

Mulvaney, K (2001) 'Cultural images: the petroglyphs of a sandstone quarry, Helen Springs, Northern Territory, Australia' 18(1) *Rock Art Research* 40–54

Mulvaney, J and Kamminga, J (1999) *Prehistory of Australia*, St Leonards: Allen and Unwin

Nganjmirra, N (1997) *Kunwinjku Spirit: Creation Stories from Western Arnhem Land*, Melbourne: Melbourne University Press

O'Connell, JF (2000) 'An emu hunt', in Anderson, A and Murray, T (eds), *Australian Archaeologist: Collected Papers in Honour of Jim Allen*, Canberra: Australian National University

O'Connor, S (1999) *30,000 Years of Aboriginal Occupation: Kimberley, North West Queensland*, Canberra: Australian National University Press

Palmer, K (1977) 'Aboriginal sites and the Fortesque River, north west of northern Australia' 13(1) *Archaeology of Physical Anthropology in Oceania* 226–33

Paton, R (1994) 'Speaking through stones: a study from northern Australia' 26 *World Archaeology* 172–84

Petrie, CC (1904) *Tom Petrie's Reminiscences of Early Queensland*, Sydney: Angus and Robertson

Postgate, N (1997) 'Mesopotamian petrology: stages in the classification of the material world' 7(2) *Cambridge Archaeological Journal* 205–24

Price, NS (2001) *The Archaeology of Shamanism*, London: Routledge

Rappaport, RA (1999) *Ritual and Religion in the Making of Humanity*, Cambridge: CUP

Robinson, R (1966) *Aboriginal Myths and Legends*, Melbourne: Sun Books

Ross, J (1997) *Painted Relationships: An Archaeological Analysis of a Distinctive Anthropomorphic Rock Art Motif in Northwest Central Queensland*, unpublished Hons thesis, University of New England

Roth, WE (1897) *Ethnological Studies Among the North-West-Central Queensland Aborigines,* Brisbane: Government Printer

Roth, WE (1903) 'Superstition, magic and medicine', *North Queensland Ethnography Bulletin No 5,* Brisbane: Government Printer

Roth, WE (1904) 'Domestic implements, arts, and manufactures', *North Queensland Ethnography Bulletin No 7,* Brisbane: Home Secretary's Department, Department of Public Lands (facsimile reprint: *The Queensland Aborigines,* Volume 2, Aboriginal Studies Series No 3, 1984, Victoria Park, WA: Hesperian Press)

Roth, WE (1907) 'Burial ceremonies and disposal of the dead', *North Queensland Ethnography Bulletin No 9, Records of the Australian Museum* 5, 365–403

Sassaman, KE (1997) 'Acquiring stone, acquiring power', paper presented at the 62nd annual meeting of the Society for American Archaeology, Nashville

Saunders, NJ (2001) 'A dark light: reflections on obsidian in Mesoamerica' 33(2) *World Archaeology* 220–36

Schrire, C (1982) *The Alligator Rivers: Prehistory and Ecology in Western Arnhem Land,* Canberra: Australian National University

Sharp, L (1952) 'Steel axes for Stone Age Australians' 11 *Human Organisation* 17–22

Sinclair, A (1995) 'The technique as symbol in Late Glacial Europe' 27(1) *World Archaeology* 50–62

Smith, ML (1999) 'The role of ordinary goods in Premodern exchange' 6(2) *Journal of Archaeological Method and Theory* 109–35

Smyth, RB (1876) [1972] *The Aborigines of Victoria with Notes Relating to the Habits of the Natives of Other Parts of Australia and Tasmania, Volume 1,* South Yarra: John Currey (republished 1972)

Spencer, B (1928) *Wanderings in Wild Australia, Volume 2,* London: Macmillan

Spencer, B (1966) *Native Tribes of the Northern Territory of Australia,* Oosterhout: Anthropological Publications

Stanner, WEH (1933) 'Ceremonial economics of the Mulluk Mulluk and Madngella Tribes of the Daly River, North Australia: a preliminary paper' 4 *Oceania* 156–75

Stanner, WEH (1960) 'On Aboriginal religion', *The Oceania Monograph* 11

Steele, JG (1983) *Aboriginal Pathways in Southeast Queensland and the Richmond River,* St Lucia: University of Queensland Press

Stout, D (2002) 'Skill and cognition in stone tool production: an ethnographic case study from Irian Jaya' 43(5) *Current Anthropology* 693–722

Taçon, PSC (1989) *From Rainbow Snakes to 'X-Ray' Fish: The Nature of the Recent Rock Painting Tradition of Western Arnhem Land, Australia,* unpublished PhD thesis, Australian National University

Taçon, PSC (1991) 'The power of stone: symbolic aspects of stone use and tool development in western Arnhem Land' 65 *Antiquity* 192–207

Taçon, PSC and Brockwell, S (1995) 'Arnhem Land prehistory in landscape, stone and paint', in Allen, J and O'Connell, JF (eds), *Pleistocene to Holocene in Australia and Papua New Guinea, Transitions: Antiquity* special no 265, 676–95

Taçon, PSC and Chippindale, C (2001) 'Transformation and depictions of the first people: animal-headed beings of Arnhem Land, NT, Australia', in Helskog, K (ed), *Theoretical Perspectives in Rock Art Research,* Oslo: Novus Forlag

Taçon, PSC and Chippindale, C (2002) 'Changing places: North Australian rock-art transformations 4000–6000 BP', paper presented at the 9th International Conference on Hunting and Gathering Societies, Heriott-Watt University, Edinburgh, Scotland (forthcoming with a Session Proceedings Volume edited by Papagianni, D and Layton, R)

Taçon, PSC, Wilson, M and Chippindale, C (1996) 'Birth of the rainbow serpent in Arnhem Land rock art and oral history' 31 *Archaeology in Oceania* 103–24

Thomson, DF (1949) *Economic Structure and the Ceremonial Exchange Cycle in Arnhem Land*, Melbourne: Macmillan

Tibbett, K (2002) 'Archaeological analysis of stone axe exchange networks in the Lake Eyre basin during the mid- to late Holocene' 55 *Australian Archaeology* 22–29

Tindale, NB (1965) 'Stone implement making amongst the Nakako, Ngadadjara and Pitjandjara of the Great Western Desert' 15 *Records of the South Australian Museum* 131–64

Tindale, NB (1974) *Aboriginal Tribes of Australia: Their Terrain, Environmental Controls, Distribution, Limits, and Proper Names*, Canberra: Australian National University

Towle, CC (1939) 'Stone arrangements and other relics of the Aborigines found on the Lower Macquarie River, NSW, at and near Mount Foster and Mount Harris' 7 *Mankind* 200–09

Van der Leeden, AC (1975) 'Thundering Gecko and Emu: mythological structuring of Nunggubuyu patrimoieties', in Hiatt, LR (ed), *Australian Aboriginal Mythology*, Canberra: Australian Institute of Aboriginal Studies

Walsh, GL (1988) *Australia's Greatest Rock Art*, Bathurst: Robert Brown and Associates

Warburton, M and Duke, P (1995) 'Projectile points as cultural symbols', in Duke, P and Wilson, MC (eds), *Beyond Subsistence: Plains Archaeology and the Postprocessual Critique*, Tuscaloosa: University of Alabama Press

Warner, WL (1969) *A Black Civilization*, Gloucester: Peter Smith (revised edn)

Welch, DM (1992) 'Kakadu Dreaming: Ancestral Beings and mythology in the rock art of the Kakadu region', in McDonald, J and Haskovec, IP (eds), *State of the Art: Regional Rock Art Studies in Australia and Melanesia*, Melbourne: Australian Rock Art Research Association

– 9 –

GEOARCHAEOLOGY AND THE GODDESS LAKSMI

RAJASTHANI INSIGHTS INTO GEOARCHAEOLOGICAL METHODS AND PREHISTORIC SOIL USE

Nicole Boivin

It goes without saying that archaeologists have always had a close association with dirt – digging through the stuff is what archaeologists do, and the good old garden trowel remains the pre-eminent tool and indeed symbol of archaeological research. However, it is only relatively recently that archaeologists have really begun to take an interest in dirt itself, and started to see it as more than just something to be shifted to get at the things of real interest. Developments in the study of dirt, as well as other elements of the mineral world such as stones, ores and mineral precipitates, have led to the establishment of a separate archaeological sub-discipline called geoarchaeology, which is aimed at exploring how the study of natural and archaeological sediments can shed light on past practices and environments, as well as the formation of archaeological sites. During its formative period, this sub-discipline has drawn heavily on the methods and theories of the natural sciences, and in particular geology and the sciences. The aim of this chapter is not so much to discount the natural sciences as an important source for geoarchaeology as to draw attention to the need for input from the methods and theories of the social sciences into the field as well, particularly in the study of occupation sediments.

The genesis of this paper can be traced back to my encounter, in rather unexpected circumstances, with the Hindu goddess Laksmi. But before embarking on a discussion of this event and my subsequent findings, I should clarify the context in which the research project to which I shall allude took place. Between July 1997 and January 1999, as part of my PhD research, I spent a total of nine months in the small, rural village of Balathal in the southern part of the state of Rajasthan in north-western India (**Figure 9.1**). During this time, I was primarily engaged in an ethnoarchaeological study of domestic space that focused in particular on the social and symbolic aspects of Balathal houses (Boivin 2001). However, due to my background in geoarchaeology (see Boivin and French 1997–98, for example), I also took an interest in the physical properties of houses and the processes involved in their formation and conversion into archaeological sites. I not only adopted the participant observation and interview methodologies of contemporary ethnography, but also examined people's houses and particularly house floors using the methods of geoarchaeology and especially soil micromorphology. This combination of interpretive and geoarchaeological methods represents an unusual approach in archaeology and was primarily responsible for the insights that are reported here.

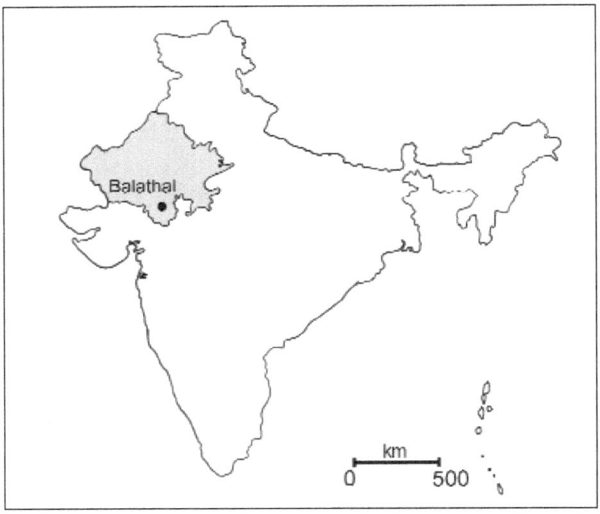

Figure 9.1: Map showing location of Balathal within India. The state of Rajasthan is shaded.

GEOARCHAEOLOGY AND LAKSMI

In using the unique approach that I did in my research, one of my aims was to draw a bridge between the rich aesthetic, social and symbolic features of rural Rajasthani houses revealed by ethnography on the one hand, and the sediments from house floors and other sources that are frequently the main focus of archaeological and particularly soil micromorphological enquiry into ancient houses and other structures on the other. To this end, I initially set out to try to find out more about the soils (*mitti* in the local dialect)[1] that villagers use to construct and maintain houses, and their reasons for using them. In doing so, it quickly became apparent to me that there were some important differences between the way that villagers described and classified soils, and the methods commonly employed by geoarchaeologists in the description and classification of soil types.

The case of *pili mitti* perhaps illustrates this point most dramatically. It was in studying *pili mitti* that I was first acquainted with the Hindu goddess Laksmi. Villagers pointed to the red-coloured soil known as *pili mitti*, and told me that it contained Laksmi, or even more frequently, that it *was* the goddess herself. *Pili mitti*, it turned out, is one of the many incarnations of Laksmi, the goddess of wealth and good fortune (**Figure 9.2**). Accordingly, *pili mitti* is highly prized in the village – at least in architectural contexts – for its auspiciousness, and is frequently used in house plastering despite the fact that it is not very resistant to trampling and weathering (**Figure 9.3 colour section**). Especially auspicious is *pili mitti* collected on Dhan Teras and Dasa Mata, festival days associated with the worship of Laksmi. Bright red *pili mitti* is also more desirable than duller-coloured varieties of this soil, although it should be pointed out that not all bright red soils are associated with Laksmi. When queried, some villagers argued that *geru*, a red soil marketed in nearby towns and cities, is not an incarnation of Laksmi. Thus for villagers, the salient features of *pili mitti* are that it is red and auspicious, due to its association with Laksmi. The soil type is further sub-divided into categories based on when and where it is collected, and how brightly coloured it is.

Figure 9.2: A 'god poster' depicting the Hindu goddess Laksmi.

This classification of *pili mitti* (a name that actually translates as 'yellow soil') contrasts sharply with descriptions of the soil according to standard soil science and geoarchaeological norms. In fact, since geoarchaeological classificatory systems emphasise soil texture over colour, *pili mitti* is not even a logical category within geoarchaeology. Villagers group under the *pili mitti* label a variety of red-coloured soils that a geoarchaeologist would separate into different categories. Brightly coloured *pili mitti* is categorised as a calcitic silt loam, while duller *pili mitti* is really more of a clay loam. For geoarchaeologists, the iron impregnation of the soil fabric that gives the soils their colour does not provide grounds for grouping the soils into a single category. Furthermore, geoarchaeologists are not interested in where or when a soil was collected; soils collected at different times from opposite sides of the planet can be considered identical from a soil science point of view if their physical properties are the same. Geoarchaeologists are more interested in the texture of a soil, and the internal arrangement of its components (its structure or fabric), as well as its pH, organic matter content, calcium carbonate content and other such scientific variables.

A similar divergence between geoarchaeological and village soil descriptive and classificatory systems can be observed when other soil types are considered. For example, *kali mitti* is defined by its black colour (*kali* means black), and is accordingly described as inauspicious or impure by villagers. Thus, while it is recognised as providing a good material for building purposes due to its strength and plasticity, its widespread use in domestic construction is masked by the common application over the top of it of plaster made from soil of another colour. Geoarchaeologists, in contrast, do not emphasise the colour of this soil so much as its texture and organic matter content. The soil would commonly be classed as a clay loam with a relatively high organic matter content.

Safed mitti provides a similar contrast. While geoarchaeologists class the soil as a calcitic loam and emphasise its calcium carbonate content, villagers once again emphasise the colour of this soil, and another property with which it is associated – its ritual purity. Extremely pure white soils are highly sought after for house plastering purposes, due in part to the links between white and purity, and may be collected from distant sources and even brought as gifts by visiting relatives. They are often mixed with water and formed into soil 'cakes' that when dry permit safe, convenient and even aesthetically-pleasing storage of the soil (**Figure 9.4**). Such properties as its beauty and ritual purity mean that *safed mitti* is commonly used in the domestic context despite the fact that its low clay content makes it poorly resistant to weathering.

What we can see from even these limited examples, then, is that the Western scientific descriptive and classificatory principles normally employed by geoarchaeologists are of a very different sort from those used by Rajasthani villagers in assessing and talking about the soils used in building houses. Though there is some overlap between the two approaches, there are nonetheless important differences. Most geoarchaeologists would see a soil classification system based primarily on colour as problematic, and one that considered the

Figure 9.4: Cakes of *safed mitti* in a sculpted house alcove.

sacredness, auspiciousness, purity, and time and place of collection of a soil as entirely inappropriate and irrelevant. Villagers, on the other hand, do not, like geoarchaeologists, classify soils primarily according to texture, even though this property is acknowledged to contribute substantially to the durability and longevity of the structural features soil is used to create. They also have little interest in standard scientific soil parameters, such as pH, organic content and calcium carbonate content.

Naturally the question that follows is, are these kinds of differences relevant? Geoarchaeologists, after all, are not usually attempting to reconstruct the soil-related perceptions of past people, so much as their soil-related practices. They are interested in how people used soils, and how this use contributed to the formation of archaeological sites as we find them today. In reconstructing these activities, what they need to focus on are not the meanings that soils hold, but the physical processes that led to their deposition, which can be linked to particular patterns of behaviour. This reconstruction of activities relies on an understanding of physical properties and uniformitarian principles that are open to objective elucidation, rather than subjective perceptions that derive from an emic perspective. While people in the past may have held ideas about soils that differ from those of geoarchaeologists today, this is not relevant to the geoarchaeological enterprise.

My argument would be that in fact this reasoning is incorrect. Contrary to the thinking of many practitioners, people's perceptions of soils are, at least in some cases, a relevant aspect of geoarchaeological research. The way geoarchaeologists understand and evaluate soils can and does impinge on how they reconstruct past activities, despite claims of objectivity and appeals to universal principles. In the remainder of this paper, by way of two key examples, I would like to demonstrate, first, that perceptions of soil are in some cases key to the ways that people use soil, and thus to site formation processes, and secondly, that soil perceptions themselves may be of interest to archaeologists.

FLOOR PLASTERING AS A MEANINGFUL PROCESS

In my first example, I will focus on the geoarchaeological analysis of occupation sediments, and in particular the soils that make up structural remains, such as the remnants of mud houses that are excavated on archaeological sites across the world. One of the tasks that geoarchaeologists have increasingly undertaken is the investigation, using the technique of soil micromorphology in particular, of the construction methods and materials, and floor deposits, of the structures unearthed in archaeological excavations. While still in a pioneering phase, such studies have nonetheless suggested that the detailed analysis of structural-related sediments is a potentially worthwhile enterprise (eg Boivin 2000; Courty *et al* 1989; Matthews *et al* 1997a; Middleton and Price 1996). I would argue that realisation of this potential will have much to do with the degree to which geoarchaeologists take on board consideration of past perceptions of the soils and other materials that are studied under the microscope and by way of other methods of chemical and structural analysis.

To illustrate my point I will return once again to Balathal village. Here, as I have already indicated, I undertook the study not only of soils specifically, but also of the wider practices and formation processes associated with the construction and maintenance of the mud houses that the majority of villagers still inhabit. What I learned in doing so is that the use of soils must be understood in the context of activities that are not only functionally, but also socially and symbolically meaningful. In taking account of such non-functional considerations, such findings go against the grain of standard accounts of traditional construction methods that are found in both the academic/specialist (eg Facey 1997; McHenry 1984) and mainstream public literature on vernacular architecture, which have tended to inspire archaeological and geoarchaeological interpretations of structural sediments (eg Adam 1994; Courty *et al* 1989; Kemp 2000; Matthews 1995). Embedded within architectural discourses that are often subtly anti-modernist in orientation, and at the very least tend to romanticise vernacular architecture by appealing to its supposed simple functionalism, such accounts inevitably argue that traditional buildings are optimally adapted to local materials and environmental conditions. My own studies in Balathal suggest that the situation is much more complex than this, at least in rural Rajasthan and probably also elsewhere in the world where mud-built architectural is, or was, the norm. This is not to say that physical, functional or environmental factors never play a role in structuring the choice of materials and techniques employed in creating mud houses, but only that other considerations also prove in many cases to be important as well.

The degree to which non-functional considerations impinge on building and maintenance activities has already been hinted at in the above discussion of soil classification in Balathal. While it is clear that Balathal villagers are aware of the fact that *kali mitti* (black soil) is a good material for building with, and often use it for creating important structural elements in houses, it is also clear that there are all sorts of situations in which they deliberately eschew the soil in favour of less durable soil types. This is particularly obvious in the context of house plastering. Balathal houses are constantly replastered, both inside and out, over the course of any given year. While some degree of regular replastering is required to prevent mud houses from decaying as a result of exposure to the elements (Carter and Pagliero 1966; Facey 1997; Rothschild *et al* 1993), it is nonetheless the case that most replastering takes place for other reasons. It is these reasons that form the context for decisions about which soil types will be used for plastering, and that ultimately affect the patterns of sedimentary deposition in Balathal houses.

Most plasters in rural Rajasthan are made using some mixture of soil and cow dung. The cow dung adds strength to the plaster mixture, and is also considered to be purifying in the Hindu tradition (Babb 1970; Khare 1962). While the amount of cow dung that is added to a plaster mixture is an important source of plaster variation, the main difference between different types of plasters derives from the types of soils used in their creation. These different types of soils confer their properties onto the plasters they are used to create. As we have seen, Balathal villagers' perceptions about what constitute the pertinent qualities of a soil vary somewhat from those of geoarchaeologists. These qualities include, for example, colour, auspiciousness, purity and date of collection.

By way of example, let us return again to *pili mitti*, the red soil that contains Laksmi. Because of the association between Laksmi and *pili mitti*, plasters made from this soil are understood to protect and bring auspiciousness to the spaces to which they are applied. Thus *pili mitti* is often used to make those plasters that are applied to houses on special occasions like weddings and festivals, and in particular sacred areas, like the *puja* (worship) area (**Figure 9.5**) and hearth (the latter can be considered a sacred space in the Hindu house; see **Figure 9.6 colour section**). Often when people talk about spaces plastered with *pili mitti*, they make it clear that they consider that the soil actually confers holiness, and even a divine presence, onto such spaces. For example, I was told that meat should not be eaten in a room where *pili mitti* has been plastered above the hearth or on the floor, because it is wrong to place impure things in the vicinity of a deity (Boivin 2001: 91).

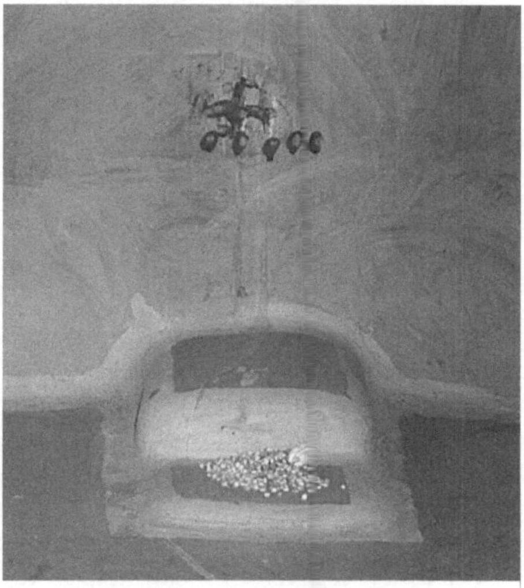

Figure 9.5: *Puja* mound in a Balathal house, with an offering of corn on its lower tier.

Plaster is applied not only to confer auspiciousness to particular places at particular moments, however, but also to purify spaces. While purity is associated with cow dung, it is also associated with red and, in particular, white soil. Black soils, meanwhile, are considered impure, and are rarely used for plastering, at least in external coats of plaster. Since houses, where cooking and *puja* (worship) as well as other important activities are performed, should be pure, lighter coloured (white or beige) soils are commonly used for external plaster coats. Coats of light coloured plaster are also applied to purify a space that has been defiled through birth, death or, at least in theory, the presence of something or someone impure (such as meat or a low caste person). Liminal phases after birth or death usually end with a major replastering event, whether of the whole house or just the room occupied by the mother and child or deceased. In such cases, *pili mitti* is not an appropriate soil to plaster with, since its sacredness means it should not make contact with impurities.

Of course plastering also has an important aesthetic and social element. Pure and auspicious soils also create aesthetically pleasing plasters, and one additional reason for repeated replastering is to ensure that families present a respectable face to the outside world. This is clear from the following quote from the Abbé Dubois' *Hindu Manners, Customs and Ceremonies*:

> Hindus purify their houses day by day from the defilements caused by promiscuous goers and comers. It is the rule amongst the upper classes to have their houses rubbed over once a day with cow-dung, but in any class it would be considered an unpardonable and gross breach of good manners to omit this ceremony when they expected friends to call or were going to receive company. (Dubois 1906: 154)

In this way, then, women, who are solely responsible for household plastering, contribute in an important way to the appearance of pure living and moral uprightness that most families aim to project (though note that women in Balathal, where servants are unheard of, are too busy with agricultural and other activities to consider replastering on a daily basis, or even as frequently as stated claims suggest they do). Women also help to generate prosperity and success by creating a domestic realm that will be welcoming for supernatural beings, as the following advice from the *Stridharmapaddhati*, an 18th-century compendium of rulings on the behaviour of a good Hindu wife, makes clear:

> The goddess Laksmi, complete with all her attributes, always dwells in the house that always shines as a result of smearing [with cow dung] and auspicious powder designs. (Cited in Leslie 1991: 112)

Finally, it should be pointed out that plasters also serve to some degree as spatial cues in rural Rajasthani houses. Different types of plaster on floors and walls signify different types of spaces, entry to which may be restricted based on gender, caste, or state of purity, for example. The use of different plaster types allow the boundaries of hearths and *puja* (worship) places, both places that should be kept pure, to be clearly delineated. They can also signal changes in the meaning of a space. Since houses in rural Rajasthan are frequently small, the same space is often used in different ways over time. Replastering can help both to create an appropriate space for a new activity, and at the same time cue its existence. For example, ritual space can be enlarged in preparation for a particular ritual activity using a plaster made of *pili mitti*.

Plastering, then, is not a simple utilitarian activity, but one that is deeply implicated in the creation and use of socially and symbolically meaningful spaces as part of daily, and inevitably strategic, practice. The deposition of sedimentary layers on Balathal house floors is closely linked to the processes by which domestic spaces in this rural village are imbued with meaning. These processes, furthermore, leave physical traces that are amenable to geoarchaeological analysis. I excavated block samples from floors in Balathal houses, and then processed and analysed them using the technique of soil micromorphology (see Boivin 2001 for details), which is also commonly applied to the analysis of floor deposits in archaeological contexts. This method revealed that the outcome of plastering events is visible as individual sedimentary layers in thin section (see **Figure 9.7 colour section**). Furthermore, the patterning of accumulated sedimentary layers across space is closely related to the ways in which different spaces within the house are used and invested with meaning. Thus, for example, sacred and ritually significant areas of the house, such as the hearth and *puja* (worship) areas, exhibited more plaster layers, as well as a higher percentage of red and white soils, than other areas (see **Figure 9.7 colour section**).

What the micromorphological analysis of Balathal house floors demonstrates is that the physical products of the symbolic and social processes I have addressed here are amenable to geoarchaeological analysis. That such physical residues withstand the effects of time, at least in certain contexts, is wonderfully illustrated by the complex floor and wall stratigraphic sequences that have been revealed at the 10,000-year-old Neolithic site of Çatalhöyük in Turkey through the application of micromorphological techniques (see Matthews *et al* 1996, 1997a). A number of particularly notable examples from Çatalhöyük display hundreds of individual plaster layers, with some sequences exhibiting complex, highly regular stratigraphic patterning (see, for example, Plate 3 in Boivin 2000). Soil micromorphology has also revealed complex floor and plaster sequences at other sites (Gé *et al* 1993; Matthews 1995; Matthews *et al* 1997b), which suggests that there is much potential for using this method to undertake detailed spatial analyses of excavated buildings.

Geoarchaeologists therefore need to take account of people's perceptions of soils. House building and floor plastering are not necessarily strictly utilitarian practices in the way that many archaeologists and geoarchaeologists suppose, and it is therefore necessary to be prepared for the possibility that the sedimentary deposits these activities leave behind will not allow straightforward interpretation according to scientific and rationalist principles. It may be necessary to entertain the notion that sedimentary patterns have symbolic origins, and it may furthermore prove useful for geoarchaeologists in some cases to depart from standard scientific understandings of soils, and be open to alternative ways of describing and classifying the sediments they observe under the microscope or using other means. If Balathal village were an archaeological site, it is clear that no valid interpretations of geoarchaeological samples taken from its house remains would be possible in the absence of an anthropologically and theoretically informed perspective. Geoarchaeological and scientific analysis of archaeological sediments from houses could not precede theoretical and interpretive analyses, but would need to be part and parcel of such studies. Culturally-produced sedimentary deposits on occupation sites are produced through meaningful activities that must be skilfully unravelled by the theoretically informed geoarchaeologist, rather than forced to coincide with pre-existing ethnocentric models and sterile scientific categories.

NEOLITHIC SOIL PERCEPTIONS

Shifting perspective somewhat, I now wish to turn from the ethnographic present to the Neolithic past, and specifically the Neolithic of the south-eastern Mediterranean region.

As I indicated earlier, I wish to demonstrate in this paper that not only are perceptions of soils relevant to the interpretation of geoarchaeological samples, but also that such perceptions may be of interest for their own sake. If soils are part of a society's culturally constructed and socially ordered understanding of the world, then it follows that they are of interest not just to geoarchaeologists, but also to archaeologists in general. This fact is perhaps nowhere better illustrated than in the prehistoric archaeology of the south-eastern Mediterranean, an area incorporating both south-eastern Europe and a substantial portion of the Near East. Here, the Neolithic marks a shift in human exploitation of soil resources that is awe-inspiring both in its scope and in the degree to which it has thus far been overlooked by archaeologists.

Sometimes referred to as the 'Cradle of Civilization', the eastern Mediterranean, and particularly its Near Eastern component, is argued to constitute the location of the earliest and perhaps thus most clearly independent shift from nomadic foraging to settled, food-producing forms of society. While internally diverse, in terms of both its chronology and cultural attributes, the region is nonetheless defined by a shared set of Neolithic plant and animal domesticates as well as material practices, including the construction of tells and houses with elaborate interiors. The Neolithic transition in this region is marked by the introduction of agriculture, ceramics and permanent settlement, traits that have indeed, rightly or wrongly, come to define the very term Neolithic.

What is less often realised (though see Stevanovic 1997 and nd, as well as Wengrow 1998) is that the Neolithic of the south-eastern Mediterranean region is also marked by another notable feature: the sudden reliance on soil as a major resource (Boivin 2001, 2002, forthcoming). Indeed, in the south-eastern Mediterranean, all of the major, Neolithic-defining traits – agriculture, ceramic production and the construction of permanent dwelling places – are in one way or another wholly reliant on soil. Thus, during the Neolithic transition in the south-eastern Mediterranean, people go from seeing soil as a minor, rather peripheral resource (the most salient indication of soil use in the preceding Palaeolithic period being the fascination with red ochre that is attested in a range of archaeological contexts over a long period of time (Velo and Kehoe 1990; Taçon, this volume), to exploiting it for as wide a range of purposes as it is perhaps ever used at any one time in human history (see also Schmandt-Besserat 1974). South-eastern Mediterranean Neolithic societies were soil-based societies, surrounded by the material products of an intensive engagement with and understanding of the natural properties of soil. Soil was the material that nurtured their crops, and from which they fabricated their built environment and created their material implements, furnishings and surroundings. Soil was used not only to form the first ceramics, but also to fashion an unprecedented array of symbols and artistic items, including jewellery, decorative items, figurines, reliefs, sculptures, paintings and plasters. Soil is, in addition, the primary component of the large settlement mounds or tells that characterise the Neolithic of this region, and that signify a fundamental and undeniably crucial shift in the relationship between people and places during this period.

As I have indicated already, a few archaeologists have noted the importance of soil to the Neolithic story in the south-eastern Mediterranean – Stevanovic, for example, has called the Neolithic 'the Age of Clay' (Stevanovic 1997). For the most part, however, the vast majority of archaeologists, whatever their disciplinary background, have overlooked the fact that soil is the common denominator in the advent of all major Neolithic traits in the south-eastern Mediterranean. This fact goes largely unrecognised. The reasons for this oversight are no doubt complex, but may at least in part be attributed to a widespread perception of soils as both mundane and non-cultural. For most Westerners, the word 'soil' conjures up, at best, images of potted plants and dark crumbs that need to be washed off mushrooms purchased from the supermarket, and the concept is only partially rescued by the somewhat more exciting synonym 'clay'. Plants and animals may have been domesticated by human societies, which radically altered and shaped them while relating to them in new ways, but soils have only ever lain mute at our feet, an endlessly renewable and unchanging resource. And yet soil clearly did change – soil in the Palaeolithic was not the same as soil in the Neolithic, and the Neolithic ushered in a change not just in the use of soil, but also in its perception. South-eastern Mediterranean Neolithic soil use is suggestive of a deep interest in and understanding of soil and its physical properties – its ability to confer life, be shaped and reshaped into complex forms and accumulate at particular localities. Something which loomed so large in people's lives, which surrounded them while they worked and slept, which fed and sheltered them, and was engaged with by them in intimate, sensory encounters on a continual, daily basis as soil was kneaded, moulded, mixed, pounded, sieved, watered, dried and fired, could not but have been incorporated into a worldview that at once recognised its importance and made sense of its place in Neolithic society.

The perception of soil by Neolithic societies in the south-eastern Mediterranean is a subject that warrants further study. Soil, given its central place in Neolithic societies, must have been part of a wider cosmology or worldview, and studying its use should therefore give us clues about that worldview. What soils were selected, and how they were used in different contexts, are questions that should be addressed through the collection of relevant data by geoarchaeologists and others. In Balathal, as I hope I have demonstrated, soil classificatory systems emphasise the importance of, amongst other properties, auspiciousness, ritual purity and sacredness, all of which are key concepts in many aspects of rural Rajasthani economic, social and religious life. It is likely that the use of soils by eastern Mediterranean societies was similarly structured by overarching cultural principles that soils can thus play a role in teasing out. At the site of Çatalhöyük in Turkey, for example, several investigators have observed that purity may have been an important consideration in soil use, at least in certain contexts. Examination of soil composition suggests that pure soils were specially selected or created for in-filling abandoned houses, creating foundation deposits, and burying the dead (eg Martin and Russell 2000; Matthews and Farid 1996). Soil categorisation and use thus appears to articulate a cultural opposition between purity and pollution, which emerges as a recurring leitmotif in various contexts at Çatalhöyük (Martin and Russell 2000; for more details, see Boivin 2001). Soils therefore need to be analysed by archaeologists as components of a wider social and cultural sphere, a caveat that is especially pertinent in societies, like those that existed in the eastern Mediterranean during the Neolithic period, in which soil played such a pivotal material role.

The relationship between soil perceptions and general worldviews is of course not a uni-directional one. Understanding how soils are perceived and used can provide insights into wider social and cultural principles, but such principles can also shed light on why soil became such a major resource during the Neolithic period in the first place. Soils are not just symbols with arbitrarily-attached meanings (see Boivin forthcoming), but physical entities with unique properties that may well have articulated and even reinforced certain cognitive, social, cultural and perceptual changes that were underway in the Neolithic period (Boivin 2002). Soil, for example, as an eminently malleable substance, enabled the rapid and efficient production of an unprecedented range of containers, enclosures and spatial divisions during the Neolithic, a property that likely made it a highly appropriate resource in a society in which social divisions were becoming increasingly manifest (Boivin 2001; see **Figure 9.8**). Wilson has emphasised the importance of enclosure in 'domesticated' societies, which he argues contrast with 'open' (foraging) societies as a result of their emphasis on boundaries, categories and compartments (Wilson 1988). Domesticated societies, like the Neolithic societies of the eastern Mediterranean, are all about order and exclusion, and soil is an ideal substance for mapping and encouraging divisions between people and things. The inherent properties of soil are crucial to understanding how it became such an attractive resource during the Neolithic period in the eastern Mediterranean.

Figure 9.8: The Neolithic site of Çatalhöyük in Turkey, where archaeologists have unearthed some of the earliest mud-built dwellings in the world (photo courtesy James Mellaart).

The malleability of soil may also have fed into a fascination with transformation in the Neolithic. The Neolithic appears to have featured an exponential increase in the degree to which things and landscapes were subjected to processes of human transformation and modification. While this is partially due to an increase in the *scale* of human transformations, it is also the result of important shifts in the *kinds* of transformations imposed on the surrounding world (Boivin 2001, 2004). For example, the Neolithic period saw in various places the development of a number of transformative technologies that, unlike the majority of those that preceded them in the Palaeolithic, were not effected primarily through physical force (as when stone is transformed into a tool), but rather through the addition of heat and water (Boivin 2001). As other archaeologists have also noted, mechanical transformations were gradually supplemented by an increasing number of chemical and structural modifications during the Neolithic period (Childe 1965; Rollefson 1990). Soil technologies in the eastern Mediterranean may be seen as part of this overall shift. Soil was mixed with water to create structural changes that allowed it to be transformed into furnishings, figurines, houses and village landscapes, and it was further heated in order to create chemical changes that resulted in lasting ceramic containers and other items. Modifications to soil that allowed it to become such a focal resource during the Neolithic derived from a growing interest in the human capacity to create substantial, durable changes to the natural world. The inherent properties of soil, which allowed substantial transformations through the application of mechanical and, in particular, non-mechanical but still basic technological procedures, articulated effectively with forms of engagement with the world that during the Neolithic focused on its modification and transformation to an unprecedented degree.

Examination of the Neolithic emphasis on soil in the eastern Mediterranean also needs to take into account soil's role in a range of new symbolic and communicative processes. It is widely recognised that the Neolithic featured an explosion in material symbols, a shift that may be seen as suggestive of a significant increase in the distances over which symbolic communication was possible, the sizes of the audiences which participated in such communication, and perhaps also the stability of symbolic meanings. This explosion no doubt also linked into the creation of an increasingly rich and expressive domestic environment (see Bailey 1990, 1996; Banning and Byrd 1987; Byrd 1994; Chapman 1991; Hodder 1990; Rollefson *et al* 1990; Stevanovic nd; Tringham 1991, 1995; Watkins 1990; Whittle 1996b; see **Figure 9.9**), a factor that likely contributed to the production of new social relationships and ways of engaging with the world during the Neolithic period (Byrd 1994; Chapman 2000; Hodder 1990; Kuijt 2000; Whittle 1996a). That soil played a role in such proceedings is clear from the myriad of new symbolic forms that were shaped out of this malleable substance (eg Atalay forthcoming; Schmandt-Besserat 1992; Tringham and Stevanovic 1990; see **Figures 9.10, 9.11**), which not only could be quickly and easily manipulated into an endless variety of forms, but could also be made to retain such human-induced contours through simple processes of drying or heating. The important role of soil in the development of communication and the expression of symbolic ideas during the transition from village to urban life in the Near East has also been highlighted by Wengrow (1998), who has emphasised the importance of soil's unique plasticity and malleability in the creation of novel symbolic forms. Stevanovic (1997) has noted, with reference to the rise of tells in south-east Europe, that soil imparts qualities of durability and visibility to the objects it is used to create. Both these studies share in common a recognition that soil, whether used on the small or large scale, presented new opportunities for the communication and representation of ideas during the Neolithic, and may have been intimately bound up with what Cauvin (1987, 2000) has referred to as the Neolithic 'revolution of symbols'.

Of course soil did not only enable new possibilities, it also imposed new constraints during the Neolithic in the eastern Mediterranean, and these too were crucial to the development of a novel worldview at this time. For example, using soil probably helped to more or less permanently link people to particular places, since items made of soil are not very transportable. Pottery constrains mobility (see also Haaland 1997), as does the creation of other items made out of soil, particularly dwellings. The cultivation of crops, another technology that is reliant on soil, also 'ties people to the land' and encourages sedentism, which probably explains why soil is often a metaphor for ownership and belonging in various cultural contexts (see Carsten and Hugh-Jones 1995 and Lovell 1998, for example). Investing in soil means investing in place. Therefore, just as we can say that settling down encouraged the use of soil as a resource, we can probably also say that, conversely, using soil as a resource encouraged people to settle down. The south-eastern Mediterranean Neolithic may not be so much about which one came first, but rather about the dialectical relationship between the

Figure 9.9: House interior at Çatalhöyük, with leopard clay mouldings, coloured clay plasters, and floor platforms (photo courtesy James Mellaart).

Figure 9.10: Clay objects from Çatalhöyük (photo courtesy Çatalhöyük Research Project).

Figure 9.11: Clay seals from Çatalhöyük (photo courtesy James Mellaart).

two factors that encouraged and subsequently reinforced sedentarisation (Boivin 2001).

The degree to which the various ways that soil was implicated in the development of new ways of living and understanding the world during the Neolithic were actively recognised by Neolithic peoples is of course difficult to elucidate. It seems likely, however, that, given the important role of soil in many of the key transformations that took place during the Neolithic in the south-eastern Mediterranean, it was not perceived in a straightforwardly functional and mundane way. Rather, soil may itself have held a myriad of interrelated meanings that drew upon those of its physical qualities that were most instrumental to its increased exploitation. Soil may have been symbolically linked, as it is in a variety of cultures today, to concepts of fertility (Barley 1994; Lane 1998; Lovell 1998; Sillar 1996; Taylor 1996), transformation (Adam 1994; Huyler 1995), and the sacred (Arnold 1971; Blier 1987; Gosselain 1999; Sillar 1996). Soil may have been understood, as it is in the Book of Genesis, as the basic substance from which humanity was created (Adam 1994: 58). This conceptualisation is shared by a number of societies (Barley 1994; Herbert 1993), including the Watchi of south-east Togo, who believe that humans return upon death to a landscape called *bome*, a place which is likened to a field of red clay where humans are remodelled and recycled back to human existence by *bomeno*, the mother of clay (Lovell 1998). While the specifics of beliefs about soil may be difficult if not impossible to reconstruct to this degree for the Neolithic, it is nonetheless important to recognise that there was likely a substantial interplay between perceptions and uses of soil, and wider cosmological notions and processes of engagement with the world at hand. Soil came to be a dominant resource in the Neolithic of the eastern Mediterranean, and understanding what this meant and how it impacted upon Neolithic society will mean investigating more than just the role of soil in subsistence and technological practices. It will also necessarily mean exploring the evocative, transformative and sensuous properties of soil, and their articulation with the unique changes that unfolded during the Neolithic period in this part of the world.

CONCLUSIONS

So, as it turns out, Rajasthani villagers do have something to teach archaeologists. Soil is neither simply a straightforward object of scientific enquiry, nor a mundane and passive medium shaped by technical processes and human design. The perception and understanding of soils is cross-culturally and chronologically variable, and, it turns out, relevant to the questions of geoarchaeologists and indeed archaeologists in general. In tracing the perception and use of soils from contemporary rural Rajasthan to the ancient eastern Mediterranean, a number of lessons have emerged. All suggest that current mainstream approaches to the mineral world within geoarchaeology and the wider discipline of archaeology are inadequate, and together they draw attention to the fact that while much of the rest of archaeology has come to recognise the importance of perception, symbolic representation and material engagement to all aspects of human life, sediments and many other components of the mineral world remain woefully mired in the backwaters of non-social archaeology. In the remaining paragraphs of this chapter I will summarise these lessons, which are crucial to any project aiming to redeem the geoarchaeological and wider archaeological study of soils and sediments.

The first lesson is of course that soils are a form of material culture. It is not enough to consider the physical properties and behaviour of soils, for they are part of the cultural world, and as such remain amenable to variable perceptions and understandings. Such perceptions impact on human engagements with soils, and thus how they are deposited on settlement sites. The study of occupation sediments and settlement formation processes must therefore take into account issues of meaning, representation, and signification. Geoarchaeologists looking down the microscope or performing chemical tests must remain open to soil qualities that lie outside the traditional sphere of geological and soil scientific enquiry. They must not model their archaeological enquiries too closely upon approaches to soils and sediments pursued by the so-called 'hard sciences'. Geoarchaeologists need to be archaeologists, and they need to recognise that for all the scientific methods they employ, ultimately, they study material culture too.

The second lesson, which follows from the first, is that soils are not just functional or technological resources. This may be phrased another way: soils are functional and technological resources, but our understanding of what this means has, in European archaeology at least, changed dramatically in the past few decades. Where once the functional and technical realms were perceived as separate from the less tangible spheres that dealt with the social, symbolic and ritual, most archaeologists now recognise not separate spheres, but an integrated whole. Functional objects are also symbolic objects (Hodder 1982; Miller 1985; Shanks and Tilley 1987), and technological processes are also social strategies (Dobres and Hoffman 1999; Lemonnier 1992; Pfaffenberger 1988, 1992). Previously clear boundary lines are now hopelessly blurred. It is necessary therefore not only to rethink how geoarchaeologists address the soils they encounter, but also how archaeology as a whole deals with soils. The dirt that remains after all the 'real' artefacts have been removed, the clay from which the piles of pot sherds were made, the collapsed mud brick that is so challenging to trace – these and other examples of the forms in which soil is encountered on archaeological sites need to be problematised and reconceived. Soil is implicated in countless mundane functional and technological processes on the settlement sites of most regions and periods. However, soil is inevitably overlooked, to a large extent invisible next to more glamorous artefacts or the finished products it is used to create. Soil that *is* analysed is studied by specialists – ceramicists, geoarchaeologists, conservationists, or archaeologists focused on architecture or terracotta figurines, for example – all of whom limit their studies to highly particular issues and questions. It is because of this state of affairs that a major 'soil revolution' like that which took place in the south-eastern Mediterranean Neolithic can go virtually unnoticed by archaeologists in one of the most (archaeologically) intensively studied areas in the world. Therefore, it is time for archaeologists to rethink how they look at soil, and to begin to both foreground soil in archaeological investigations and discussions, and remove the inter-disciplinary barriers that prevent an integrated understanding of soil use on archaeological sites. It is time to re-evaluate what is probably the most plentiful, and paradoxically at the same time most obscure, archaeological resource of all.

The third and final lesson I would like to emphasise is that soils are alive and in some senses divine. While it might seem easy to dismiss these Rajasthani understandings of soil (shared, incidentally, by a wide range of other cultures) as interesting examples of soil perception, it is worth considering for a moment how much they actually overlap with more recent academic understandings of material culture. In recognising soils as alive and divine, Rajasthani villagers attribute a certain agency to soils that is not entirely incompatible with notions of material agency that have recently been foregrounded in archaeology and anthropology (Boivin forthcoming; Gell 1998; Gosden 2001; Jones 2002). My analysis of Neolithic soil use in the south-eastern Mediterranean outlined various ways in which the specific properties of soil both constrained and enabled particular types of human behaviour and social interaction. Soil was not a passive player in symbolic, social and cognitive transformations, representing changes taking place in more abstract realms, but a key, integrated participant with an indispensable role in such changes. The physical properties of soil were integral to this role, and these belong to soil itself. They are not the product of human design, nor are they entirely at the mercy of human operators. Human beings can shape soil into a myriad of forms, but the results are not always predictable. Interacting with soil had unintended consequences in the Neolithic, moulding humans in ways they probably did not entirely expect. Interacting with soil enabled new forms of cuisine and food storage, new methods of creating social divisions, and new ways of communicating and relating to place, while at the same time imposing constraints on both short- and long-term mobility. All of these conditions eventually led to substantial, irreversible changes in human society that we continue to feel the effects of today. Humans beings may therefore have shaped soil, but soil also shaped human beings. The divine agency attributed to soil by Rajasthani villagers appears not so unlikely after all. Instead, it offers crucial insights into the relationship between human society and the material world that archaeologists and other social scientists are only really beginning to recognise.

Note

1 In this paper, the term soil is used in a general, non-scientific sense to refer to that category of matter found on the upper surface of the earth and consisting of disintegrated rock usually with an admixture of organic remains (*Oxford Dictionary*). The term thus incorporates a broad range of materials – including clay, earth, mud, adobe, loam, dirt, soil and sediment – that are sometimes differentiated for specific purposes. This use of soil derives in part from parallels drawn to the Hindi/Mewari word *mitti*, which encompasses the soil in a farmer's field, the sediment collected to build houses and plaster floors, and the clay that is the main ingredient in pottery. It should be noted that use of the term 'soil' in the paper

conflicts with standard geoarchaeological definitions, which differentiate soils from sediments. This difference is acknowledged but considered not particularly relevant to the arguments made in the present paper.

Acknowledgments

Some of the arguments made here were originally presented at a TAG session organised by Andy Jones in Oxford, and I thank both him and Bryan Hanks for their support and encouragement, as well as those who participated in the session for their useful comments. I would also like to thank those who provided critical feedback on my original paper at the 2002 EAA meeting, especially Bob Chapman. The editorial assistance of Mary Ann Owoc during the preparation of this chapter is also gratefully acknowledged, as is the assistance of Shahina Farid, who helped to locate appropriate photos from the Çatalhöyük Research Project archives. I would also like to thank Ian Hodder for granting permission to use the Çatalhöyük Project archives, and Jamie Hampson and Jinu Koshy for tracking down the Laksmi image for me in a Nandyal shop. The fieldwork discussed here was funded by the Royal Anthropological Society of Great Britain and Ireland, the Nehru Trust for the Victoria and Albert Museum, and the Smuts Memorial Fund, and would not have been possible without the assistance of Dilip Chakrabarti, Charly French, Rima Hooja, Jheevan Kharakwal, Paul Lane, Vasante Shinde, Rabindra Nath Vyas, countless other colleagues in England and India, and in particular my good friend Lorraine Fernandes and the generous inhabitants of Balathal village.

References

Adam, J-P (1994) *Roman Building: Materials and Techniques*, Mathews, A (trans), London: Batsford

Arnold, DE (1971) 'Ethnomineralogy of Ticul, Yucatan potters: etics and emics' 36(1) *American Antiquity* 20–40

Atalay, S (forthcoming) 'Domesticating clay: the role of clay balls, mini balls and geometric objects in daily life at Çatalhöyük', in Hodder, I (ed), *Changing Materialities at Çatalhöyük: Reports from the 1995–99 Seasons*, Cambridge: McDonald Institute Monographs/British Institute of Archaeology at Ankara

Babb, LA (1970) 'The food of the Gods in Chhatisgarh: some structural features of Hindu ritual' 26 *Southwestern Journal of Anthropology* 287–304

Bailey, DW (1990) 'The living house: signifying continuity', in Samson, R, *The Social Archaeology of Houses*, Edinburgh: Edinburgh University Press

Bailey, DW (1996) 'The life, times and works of House 59, Tell Ovcharovo, Bulgaria', in Darvill, T and Thomas, J (eds), *Neolithic Houses in Northwest Europe and Beyond*, Oxford: Oxbow

Banning, EB and Byrd, B (1987) 'Houses and the changing residential unit: domestic architecture at PPNB 'Ain Ghazal, Jordan' 53 *Proceedings of the Prehistoric Society* 309–25

Barley, N (1994) *Smashing Pots: Feats of Clay from Africa*, London: British Museum Press

Blier, SP (1987) *The Anatomy of Architecture: Ontology and Metaphor in Batammaliba Architectural Expression*, Cambridge: CUP

Boivin, N (2000) 'Life rhythms and floor sequences: excavating time in rural Rajasthan and Neolithic Çatalhöyük' 31(3) *World Archaeology* 367–88

Boivin, N (2001) *'Archaeological Science as Anthropology': Time, Space and Materiality in Rural India and the Ancient Past*, unpublished PhD thesis, Department of Archaeology, University of Cambridge

Boivin, N (2002) 'Dirty secrets: the untold story of Neolithic soil use', Paper presented to the Trinity College Archaeological Society, Carmarthen, Wales, 7 March

Boivin, N (2004) 'Landscape and cosmology in the South Indian Neolithic: new perspectives on the Deccan ashmounds' *Cambridge Archaeological Journal*

Boivin, N (forthcoming) 'Mind over matter? Collapsing the mind-matter dichotomy in material culture studies', in DeMarrais, E, Renfrew, C and Gosden, C (eds), *Rethinking Materiality: The Engagement of Mind with the Material World*, Cambridge: McDonald Institute Monograph Series

Boivin, N and French, CAI (1997–98) 'New questions and answers: thin section micromorphology analysis at the Souks site, Beirut' 43 *Berytus* 181–209

Byrd, BF (1994) 'Public and private, domestic and corporate: the emergence of the Southwest Asian village' 59(4) *American Antiquity* 639–66

Carsten, J and Hugh-Jones, S (1995) 'Introduction', in Carsten, J and Hugh-Jones, S, *About the House: Levi-Strauss and Beyond*, Cambridge: CUP

Carter, TH and Pagliero, R (1966) 'Notes on mud-brick preservation' 22 *Sumer* 65–75

Cauvin, J (1987) 'L'apparition des premières divinités' 194 *La Recherche* 1472–80

Cauvin, J (2000) 'The symbolic foundations of the Neolithic revolution in the Near East', in Kuijt, I (ed), *Life in Neolithic Farming Communities: Social Organization, Identity, and Differentiation*, London: Kluwer Academic/Plenum

Chapman, J (1991) 'The early Balkan village', in Gron, O, Engelstad, E and Lindblom, I (eds), *Social Space: Human Spatial Behaviour in Dwellings and Settlements*, Odense: Odense UP

Chapman, J (2000) *Fragmentation in Archaeology: People, Places and Broken Objects in the Prehistory of Southeast Europe*, London: Routledge

Childe, VG (1965) *Man Makes Himself*, Suffolk: Chaucer

Courty, M-A, Goldberg, P and Macphail, R (1989) *Soils and Micromorphology in Archaeology*, Cambridge: CUP

Dobres, M-A and Hoffman, CR (eds) (1999) *The Social Dynamics of Technology: Practice, Politics and World Views*, Washington, DC: Smithsonian Institution Press

Dubois, AJA (1906) *Hindu Manners, Customs and Ceremonies*, Oxford: Clarendon

Facey, W (1997) *Back to Earth: Adobe Building in Saudia Arabia*, London: Al-Turath

Gé, T, Courty, M-A, Matthews, W and Wattez, J (1993) 'Sedimentary formation processes of occupation surfaces', in Goldberg, P, Nash, DT and Petraglia, M (eds), *Formation Processes in Archaeological Context*, Madison, Wisconsin: Prehistory Press

Gell, A (1998) *Art and Agency: An Anthropological Theory*, Oxford: OUP

Gosden, C (2001) 'Making sense: archaeology and aesthetics' 33(2) *World Archaeology* 163–67

Gosselain, OP (1999) 'In pots we trust: the processing of clay and symbols in sub-Saharan Africa' 4(2) *Journal of Material Culture* 205–30

Haaland, R (1997) 'Emergence of sedentism: new ways of living, new ways of symbolizing' 71 *Antiquity* 374–85

Herbert, E (1993) *Iron, Gender and Power: Rituals of Transformation in African Societies*, Bloomington and Indianapolis: Indiana University Press

Hodder, I (1982) *Symbols in Action: Ethnoarchaeological Studies of Material Culture*, Cambridge: CUP

Hodder, I (1990) *The Domestication of Europe: Structure and Contingency in Neolithic Societies*, Oxford: Basil Blackwell

Huyler, S (1995) 'Clay, sacred and sublime: terracotta in India', in Fisher, N (ed), *Mud, Mirror and Thread: Folk Traditions of Rural India*, Ahmedabad: Mapin

Jones, A (2002) *Archaeological Theory and Scientific Practice*, Cambridge: CUP

Kemp, B (2000) 'Soil (including mud-brick architecture)', in Nicholas, PT and Shaw, I (eds), *Ancient Egyptian Materials and Technology*, Cambridge: CUP

Khare, RS (1962) 'Ritual purity and pollution in relation to domestic sanitation' 15(2) *The Eastern Anthropologist* 125–39

Kuijt, I (2000) *Life in Neolithic Farming Communities: Social Organization, Identity, and Differentiation*, London: Kluwer Academic/Plenum

Lane, P (1998) 'Engendered spaces and bodily practices in the Iron Age of southern Africa' in Kent, S (ed), *Gender in African Archaeology*, Walnut Creek, CA: AltaMira

Lemonnier, P (1992) *Elements for an Anthropology of Technology*, Ann Arbor: University of Michigan

Leslie, J (1991) '*Sri* and *Jyestha*: ambivalent role models for women', in *Roles and Rituals for Hindu Women*, London: Pinter

Lovell, N (1998) 'Wild gods, containing wombs and moving pots: emplacement and transience in Watchi belonging', in *Locality and Belonging*, London: Routledge

McHenry, PG (1984) *Adobe and Rammed Earth Buildings: Design and Construction*, New York: Wiley

Martin, L and Russell, N (2000) 'Trashing rubbish', in Hodder, I (ed), *Towards Reflexive Method in Archaeology: The Example of Çatalhöyük*, Cambridge: McDonald Institute for Archaeological Research/British Institute of Archaeology at Ankara

Matthews, W (1995) 'Micromorphological characterisation and interpretation of occupation deposits and microstratigraphic sequences at Abu Salabikh, southern Iraq', in Barham, A and Macphail, RI (eds), *Archaeological Sediments and Soils: Analysis, Interpretation and Managment*, London: Archetype

Matthews, W and Farid, S (1996) 'Exploring the 1960s' surface: the stratigraphy of Çatalhöyük', in Hodder, I (ed), *On the Surface: Çatalhöyük 1993–95*, Cambridge: McDonald Institute for Archaeological Research/British Institute of Archaeology at Ankara

Matthews, W, French, C, Lawrence, T and Cutler, D (1996) 'Multiple surfaces: the micromorphology', in Hodder, I (ed), *On the Surface: Çatalhöyük 1993–95*, Cambridge: McDonald Institute for Archaeological Research/British Institute of Archaeology at Ankara

Matthews, W, French, C, Lawrence, T, Cutler, D and Jones, MK (1997a) 'Microstratigraphic traces of site formation processes and human activities' 29(2) *World Archaeology* 281–308

Matthews, W, French, C, Lawrence, T, Cutler, D and Jones, MK (1997b) 'The activities inside the temple: the evidence of microstratigraphy', in Crawford, H, Killick, R and Moon, J, *The Dilmun Temple at Saar*, London: Kegan Paul International

Middleton, WD and Price, TD (1996) 'Identification of activity areas by multi-element characterisation of sediments from modern and archaeological house floors using inductively coupled plasma-atomic emission spectroscopy' 23 *Journal of Archaeological Science* 673–87

Miller, D (1985) *Artefacts as Categories: A Study of Ceramic Variability in Central India*, Cambridge: CUP

Pfaffenberger, B (1988) 'Fetishised objects and humanised culture: towards an anthropology of technology' 23 *Man (NS)* 236–52

Pfaffenberger, B (1992) 'Social anthropology of technology' 21 *Annual Review of Anthropology* 491–516

Rollefson, GO (1990) 'The uses of plaster at Neolithic 'Ain Ghazal, Jordan' 4 *Archeomaterials* 33–54

Rollefson, GO, Kafafi, Z and Simmons, A (1990) 'The Neolithic village of 'Ain Ghazal, Jordan: preliminary report on the 1988 season' 27 *Bulletin of the American Schools of Oriental Research Supplement* 97–118

Rothschild, NA, Mills, BJ, Ferguson, TJ and Dublin, S (1993) 'Abandonment at Zuni farming villages', in Cameron CM and Tomka, SA (eds), *Abandonment of Settlements and Regions: Ethnoarchaeological and Archaeological Approaches*, Cambridge: CUP

Schmandt-Besserat, D (1974) 'The use of clay before pottery in the Zagros' 16(2) *Expedition* 11–17

Schmandt-Besserat, D (1992) *Before Writing*, Austin: University of Texas Press

Shanks, M and Tilley, C (1987) *Social Theory and Archaeology*, Oxford: Polity

Sillar, B (1996) 'The dead and the drying: techniques for transforming people and things in the Andes' 1(3) *Journal of Material Culture* 259–89

Stevanovic, M (1997) 'The age of clay: the social dynamics of house destruction' 16 *Journal of Anthropological Archaeology* 334–95

Stevanovic, M (nd) 'The role of clay in shaping the Neolithic world in Southeast Europe', unpublished manuscript

Taylor, T (1996) *The Prehistory of Sex: Four Million Years of Human Sexual Culture*, London: Fourth Estate

Tringham, R (1991) 'Households with faces: the challenge of gender in prehistoric architectural remains', in Gero, JM and Conkey, MW (eds), *Engendering Archaeology: Women and Prehistory*, Oxford: Basil Blackwell

Tringham, R (1995) 'Archaeological houses, households, housework and the home', in Benjamin, DN and Stea, D, *The Home: Words, Interpretations, Meanings and Environments*, Aldershot: Avebury

Tringham, R and Stevanovic, M (1990) 'The nonceramic uses of clay', in Tringham, R and Krstic, D (eds), *Selevac: A Neolithic Village in Yugoslavia*, Los Angeles: Institute of Archaeology

Velo, J and Kehoe, AB (1990) 'Red ochre in the Paleolithic', in Foster, ML and Botscharow, LJ (eds), *The Life of Symbols*, Boulder, CO: Westview

Watkins, T (1990) 'The origins of house and home?' 21(3) *World Archaeology* 336–47

Wengrow, D (1998) '"The changing face of clay": continuity and change in the transition from village to urban life in the Near East' 72 *Antiquity* 783–95

Whittle, A (1996a) *Europe in the Neolithic: The Creation of New Worlds*, Cambridge: CUP

Whittle, A (1996b) 'Houses in context: buildings as process', in Darvill, T and Thomas, J (eds), *Neolithic Houses in Northwest Europe and Beyond*, Oxford: Oxbow

Wilson, PJ (1988) *The Domestication of the Human Species*, London: Yale UP

– 10 –

CHOOSING STONES, REMEMBERING PLACES

GEOLOGY AND INTENTION IN THE MEGALITHIC MONUMENTS OF WESTERN EUROPE

Chris Scarre

Studies of megalithic monuments in Western Europe have frequently remarked upon the varied sources of the stones used to build these structures. In the majority of cases, it is true, the materials were obtained within the local area, either from rock outcrops or (in parts of northern Europe) from the fields of glacial erratics left by retreating ice sheets. Yet the deployment of materials can rarely be explained solely by their local availability. Many megalithic structures appear by intention to incorporate stones of different geological types and origins. This is demonstrated by the fact that the different stones are often arranged not haphazardly but in specific patterns within the monuments. Symmetries, pairings and oppositions are often observed. Hence factors other than mere availability of materials must have played a part in guiding the selection of the stones, and the manner in which they were arranged within the monuments.

Varieties of stone are distinguished by a constellation of qualities both visual and tactile: their colour, their texture, their shapes and fracture planes, and by the various sizes and kinds of inclusions that they contain. Of these characteristics, it is perhaps texture that is the most apparent to the bodily encounter with the monument: passageways may be low and narrow, and brushing against the stones or feeling one's way with one's hands is an unavoidable experience in entering many of these structures. The opposition of rough- and smooth-surfaced elements has been documented at a number of sites (Cummings 2002). Yet colour, too, is often significant, and has been highlighted in a number of recent studies. It is indeed possible that many megalithic tombs originally bore painted motifs. The painted dolmens of Iberia were given prominence in the publications of the Abbé Breuil in the 1930s, and are now known not only in Galicia and northern Portugal but also in central Spain and Andalucia (Shee Twohig 1981; Devignes 1996). There are indications that painted decoration may originally have been more widely distributed in Atlantic Europe: in the rock-cut hypogea of the Marne valley (Shee Twohig 1981: 84); in Orkney passage graves (Bradley *et al* 2001); and in the *'dolmens angoumoisins'* of western France (Devignes 1996: 19).

The original geographical extent of painted decoration remains unclear despite these tantalising glimpses. The colours of the stones themselves, however, have been discussed in a number of recent studies on megalithic monuments in Britain, Ireland and Scandinavia (Jones 1999; Trevarthen 2000; Tilley 1996: 316). In each of these cases, stones of different colours are deployed in patterned ways in passages, chambers or kerbs. There remains the difficulty, however, of demonstrating the significance behind the patterning. The incorporation of differently coloured materials might be taken to suggest a symbolism of the particular colours involved, among which red, white and black are especially prominent. Ethnography has underlined the symbolic importance of red, black and white in the colour classifications of several societies, including the Yoruba of Nigeria and the Ndembu of north-west Zambia (Drewal and Mason 1998; Turner 1967). It must be emphasised, however, that colour is only one aspect of the materiality of the stones used in megalithic monuments. The use of large and often unworked slabs in itself suggests that other qualities were important, including perhaps the specific provenances of the stones, taken as they are from different places within the local landscape. It thus becomes possible to argue that in using these materials, the prehistoric builders were consciously evoking or invoking qualities, powers, or mythical events associated with particular places or outcrops.

North-western France provides several good examples of the selection and deployment of stones of different geological origin and varied colour in megalithic monuments (Scarre 2002a, 2002b). It also offers what must be among the earliest observations of colour variation among megalithic blocks. The Baron de Wismes in 1876, closely followed by Pitre de Lisle du Dréneuc in 1882, was struck by colour differences in a small group of passage graves on the northern edge of Pornic, a coastal town in the *département* of Loire-Atlantique, south of the Loire estuary. As we shall see, both authors clearly appreciated the special significance of the way in which the coloured stones were used in these monuments. It was not so much the presence of different materials; that on its own might be attributable simply to the diversity of materials that were locally available. It was the *patterning* of these materials within the monuments that gave grounds for the belief that this arrangement had been intentional on the part of the builders. Yet while what impressed the early French investigators was the colour of the stones, this in itself does not resolve the question of meaning. Did the prehistoric builders choose these particular stones *because of* their colours, or was that merely an accident or by-product of a selection of materials driven by quite different considerations?

THE PORNIC PASSAGE GRAVES

The megalithic chambers described by De Wismes and Lisle du Dréneuc were located within a row of three large Neolithic mounds on the edge of higher ground overlooking the valley and estuary of the Haute-Perche. The Tumulus de la Motte is flanked on one side by the cairn of Les Mousseaux and on the other by the monument excavated by De Wismes and known variously as the Tumulus des Six Dolmens and Tumulus des Trois Squelettes (**Figure 10.1**). The present coast follows the line of a geological fault and is marked by a steep cliff, though during the earlier Neolithic period the sea edge would have lain considerably further to the south. East of Pornic, a number of megalithic tombs line the cliff-top, though those nearest (notably the dolmens of Gourmalon and Les Hautes Folies) have been destroyed by the growth of the town (Tessier 1994; L'Helgouach *et al* 1989).

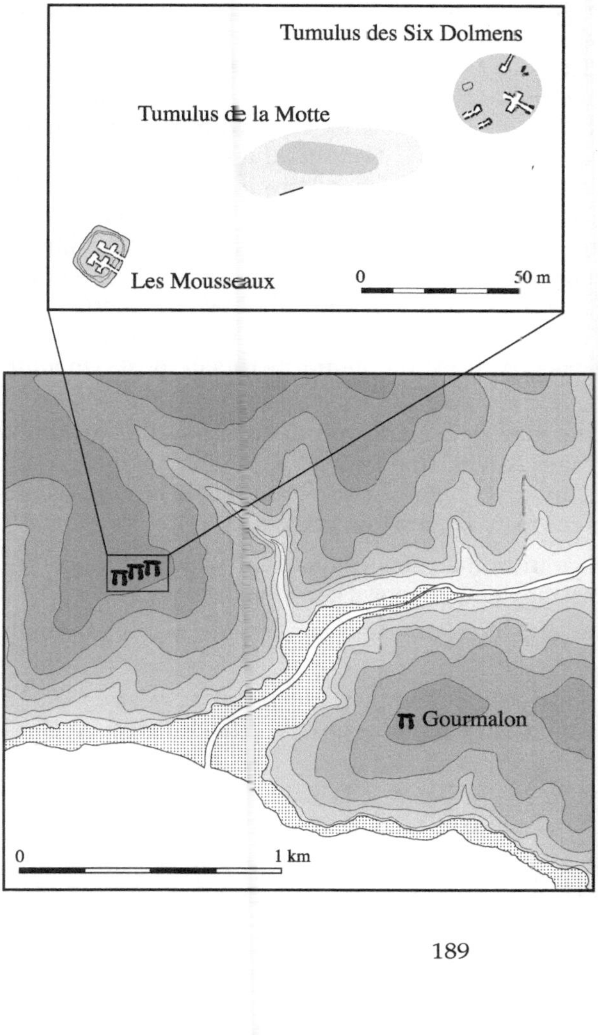

Figure 10.1: The Neolithic chambered cairns of Pornic. Contours at 5 m intervals.

It was in 1876 that the Baron de Wismes excavated a group of megalithic chambers in the Tumulus des Six Dolmens. He began by exploring the partially exposed remains of the western passage grave that he named the Caveau de la Croix (Tomb of the Cross) on account of its cruciform ground plan. The different types of stone immediately drew his attention: 'The monument consists of stones seven to eight feet tall, very carefully selected, with almost flat faces, and of excellent material, white quartz, white sandstone or ferruginous sandstone, creating one of the most beautiful constructions that Celtic monuments have ever offered' (De Wismes 1876: 208). In the caption to the lithograph illustration of this chamber, further details are provided about the placement of these coloured stones: 'To left and right of the entrance are two blocks of white quartz; they are followed by two stones of ferruginous sandstone; and these in turn by two stones of fine-grained slightly yellowish white sandstone – This curious, beautiful and intelligent symmetrical arrangement, in the pairing of the colours of the orthostats, we at least believe to be very rare' (*ibid*: 265).

Of the second passage grave that he excavated, De Wismes noted the quartz capstone at the entrance to the chamber, and the roofing of the chamber in 'stones of less dense material in a white sandstone of which none of the locals know the source' (*ibid*: 14). He later (p 229) suggests that the white sandstone came from Noirmoutier, while the ferruginous sandstone and the quartz were local to the Pornic coast, but Lisle du Dréneuc took issue with the Noirmoutier source (1882: 268). What is clear, in any event, is that the stone types had not been brought from a great distance but came from relatively local sources within a compass of 10 to 15 km. It was De Wismes' discovery of the remains of three individuals (adult male, adult female, and child) in this chamber that led him to name the monument as a whole the Tumulus des Trois Squelettes (Tomb of the Three Skeletons), while others reserve that name for this specific chamber. De Wismes went on to discover a third chamber parallel to the Trois Squelettes, then remains of a further two or perhaps three on the northern side of the mound, towards the road.

De Wismes' interest in the materials employed to construct this group of passage graves led him to commission a report from a mining engineer, Paul Poirier, which appears as an annex to his main account (Poirier in De Wismes 1876: 252–55). Poirier divides the material into four categories:

- white quartz, which comes from the veins of quartz that pass through the schist and gneiss which are the dominant local geology;
- red ferruginous sandstone, of which Poirier notes several local outcrops both on the coast and further inland;
- white and yellowish sandstone, which were popularly sourced to Noirmoutier, but which could have come from a mainland deposit such as that at Remouillé; and
- micaschist ('occasionally large slabs of micaschist'), of which Poirier says nothing further but which must also be of local origin.

The monuments were described a few years later in the *Dictionnaire Archéologique de la Loire Inférieure* (1882) by Lisle du Dréneuc, then Vice-President of the Société d'Archéologie de Nantes. He too laid particular emphasis on the kinds of stone and the manner in which they had been deployed: 'The Pornic megaliths are almost all of quartzitic sandstone or quartzite, a rock that can be found in the form of large blocks almost everywhere throughout the Pays de Retz ... It should be noted that when materials of different kinds are present in a dolmen, care has been taken to place the blocks of similar material opposite each other. In several tombs of the Tumulus de la Motte we see, among the rows of sandstone orthostats, pairings of quartz blocks: thus in the Tomb of Les Trois Squelettes, the entrance is formed by three blocks of quartz, two on either side supporting a lintel that is also of white quartz' (Lisle du Dréneuc 1882: 268; note that the 'several tombs of the Tumulus de la Motte' is in fact a reference to the Tumulus des Six Dolmens).

In 1892 Lisle du Dréneuc followed De Wismes by undertaking his own excavations in the Tumulus des Six Dolmens, discovering another (seventh) megalithic chamber (for this identification see Daniel 1939: 153; L'Helgouach 1977: 163). In this tomb, again, the colours of the orthostats had been carefully arranged: 'The ten orthostats which form this tomb are arranged in a curious way: it appears that care has been taken to alternate them depending on the nature of the stone. Thus on each side of the chamber are two orthostats of micaschist; then come two blocks of yellow sandstone, smooth-faced, forming the rear wall of the dolmen. The second orthostat on the northern side is sandwiched between two pillars of white quartz; directly opposite, a purplish orthostat of ferruginous conglomerate is followed by an orthostat of white sandstone. In this alternation of colours there is evidently an intentional scheme' (Lisle du Dréneuc 1892). What is notable once again is the systematic and deliberate arrangement of the stones of different colour and composition.

The three Pornic monuments have fared very differently since their 19th-century discovery and exploration.

In the Tumulus des Six Dolmens, two of the five chambers excavated by De Wismes were filled in immediately to ensure their preservation. Thus only three are visible for further research today, in the garden of a private house. The colour contrasts recorded by De Wismes and noted also by Lisle du Dréneuc are clearly evident, despite patination of the stones through exposure to the elements, and lichen cover. Carvings were noted by De Wismes on one of the orthostats of the Caveau de la Croix and these have recently been cleaned for study and recorded afresh. The new drawing differs from earlier versions in revealing the existence of two additional hooked or S-shaped motifs at the upper right hand side of the slab, at least one of them, and perhaps both, truncated by a break (Cassen *et al* 2001). The location of the seventh chamber excavated by Lisle du Dréneuc in 1892 remains unclear. Little survives of the original covering mound in which these chambers were incorporated.

The central monument, the Tumulus de la Motte, is the least-studied of the three, and the structures it contains remain unknown. This situation has not been improved by the construction of a windmill on top of the mound in the 18th century; nor by additional house-building in recent decades, preceded by only limited excavation that however revealed part of a kerb running at an oblique angle (L'Helgouach 1979: 559–60); nor by the more extensive destruction that was underway in 2001.

Colour and geology at Les Mousseaux

The westernmost of the three monuments, the cairn of Les Mousseaux, has survived altogether better. Initially explored in the 1840s, it was re-excavated by Jean L'Helgouach in the 1970s and subsequently consolidated and restored (L'Helgouach 1977; L'Helgouach and Poulain 1984). The cairn contains a pair of passage graves with transepted chambers, though the northernmost (probably an addition to the original monument) is asymmetrical in plan and has no transepts on its southern side. For both Daniel (1939) and L'Helgouach (L'Helgouach and Poulain 1984) the primary interest of Les Mousseaux evidently lay in its transepted form. No mention of stone colour appears in Daniel's 1939 article, or in the published report on Les Mousseaux (L'Helgouach and Poulain 1984). The omission may to some extent be excused. The colour contrasts are certainly much less striking than in the passages and chambers of Les Trois Squelettes. They are nonetheless a feature of the monument that deserves fuller attention than it has hitherto received.

The cairn of Les Mousseaux (**Figure 10.2**) is publicly signposted as one of the tourist attractions of Pornic. The monument is relatively intact and the orthostats of the two transepted passage graves are in much the same positions as when the site was first cleared by Verger in 1840. Surface patination nonetheless masks the colour of the stones. L'Helgouach describes the material as follows: 'The two dolmens are built from blocks of tertiary sandstone that can be found scattered on the eroded plateau of the pays de Retz. The spaces between the orthostats are filled by walling of micaschist slabs. The capstones do not rest directly on the

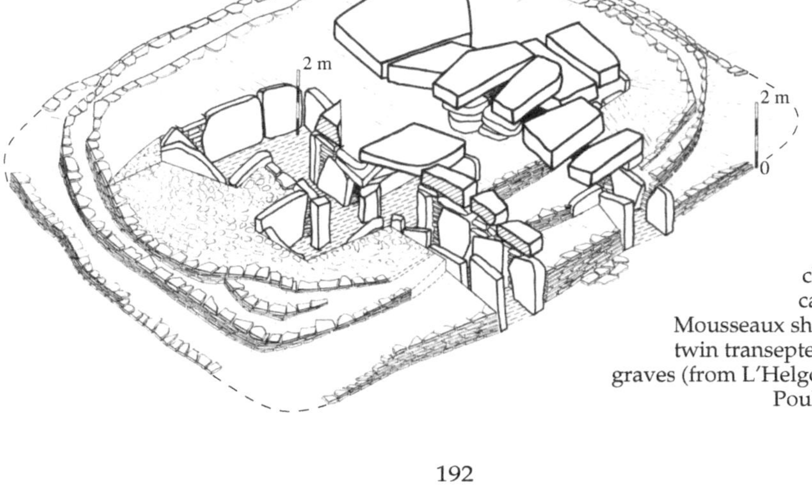

Figure 10.2: The chambered cairn of Les Mousseaux showing the twin transepted passage graves (from L'Helgouach and Poulain 1984).

orthostats but on blocks of sandstone or quartz, often of some size, that are placed between the orthostats and the capstones' (L'Helgouach 1977: 165) (**Figure 10.3 colour section**). The account of the reconstruction work makes clear that, although at least one of the capstones is a new addition, the orthostats are all original and in position, save for re-erection or straightening in some cases (L'Helgouach and Poulain 1984: note 5, p 31). Several of the sandstones are characterised by quartz inclusions, some fine-grained, some coarse-grained. The slabs themselves are generally irregular and rough-textured, though a few fine-grained, relatively smooth slabs of reddish stone are found, while at the other extreme are some with a particularly pitted and knobbly surface texture. These readily suggest carvings or engravings, but careful inspection invites caution. The only confirmed carvings are a series of cup-marks that follow the edge of one of the orthostats of the northern passage (L'Helgouach 1977: 165).

The orthostats at Les Mousseaux are of tertiary sandstone, and the colours are at first very difficult to discern, the predominant initial impression being of similar-coloured greys. Where the stones are less weathered, however, or where they are broken, the colour contrasts are striking, with reds and red-browns alongside yellow and the predominant grey. This may suggest that these slabs were unweathered blocks when first used. A number of other visual and tactile characteristics also stand out: surfaces that are smooth and fine-grained; surfaces marked by ripple or wave patterns; surfaces deeply and regularly pitted to give a kind of 'egg-box' effect; and the presence (both visual and textural) of quartz and mica inclusions of varying size and prominence in the different slabs. These qualities must have been equally apparent to the prehistoric communities who constructed these monuments. Furthermore, the pattern of colouration already remarked upon at La Motte and Les Trois Squelettes indicates that the deployment of different kinds and qualities of stone is unlikely to have been at random.

At Les Mousseaux, colour patterning is clearest in the passages (**Figure 10.4**). The northern passage grave has three pairs of orthostats in its outer passage: the outermost pair of grey and grey-yellow sandstone respectively, the intermediate pair of red ferruginous sandstone, and the innermost pair again of grey and grey-yellow sandstone, the innermost slab on the northern side covered by deep pitting so as to produce an 'egg-box' effect. In the southern passage grave, the outermost pair in the outer passage are both of grey sandstone but they present a strongly contrasting appearance, the southern orthostat being medium-grained, while the surface of the northern orthostat is entirely covered with deep pits. The intermediate pair of orthostats, like those of the northern passage grave, are of red ferruginous sandstone, and the innermost pair are of sandstone with fine quartz inclusions.

Symmetries of colour extend beyond the outer passages to the transepted chambers, the distribution of red sandstone orthostats being particularly notable (**Figure 10.5**). The northern passage grave has only a single transept, its walls formed by five slabs, two on each side and one at the back. The side slabs are paired: the first two are similar in shape, and of grey-white colour, that on the right with more conspicuous quartz inclusions, that on the left with hollows and

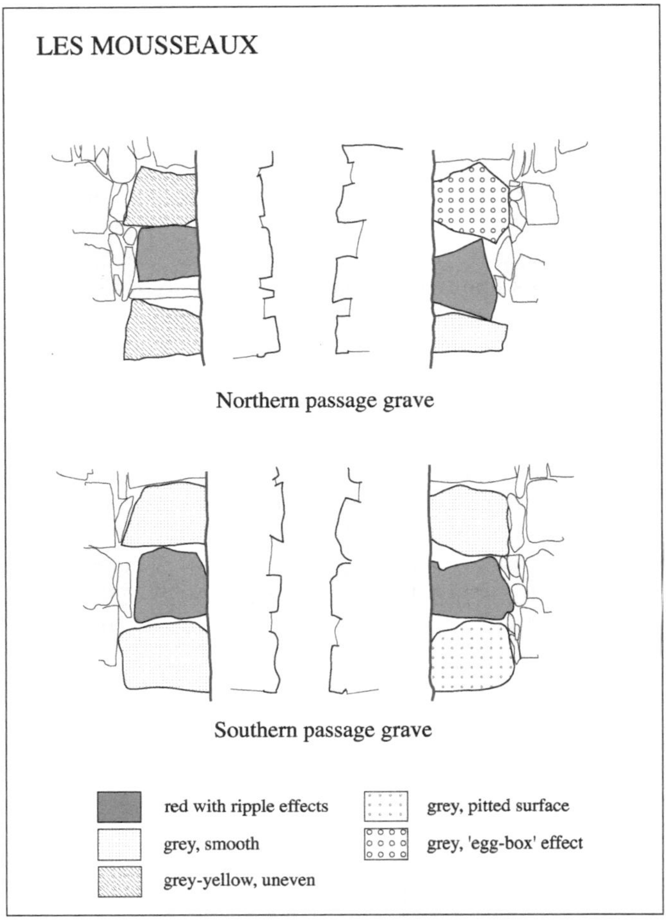

Figure 10.4: Les Mousseaux: stone colours in the outer passages of northern and southern passage graves.

voids (and some quartz); the second pair are fine-grained reddish slabs, with quartz inclusions, and 'wave patterned' surfaces; the end-slab is yellow/brown or grey, but is notably fine-grained, with no (or very few) quartz inclusions. Here we may note not only careful pairing of the stones but also a progression from rough to smooth as the rear wall of the chamber is approached.

The transepts of the southern passage grave also display a symmetrical arrangement of materials. In the northern transept, the side slabs are of identical grey sandstone with conspicuous quartz inclusions; the end-stone has pronounced hollows 20–30 cm across, and smaller voids – a smooth-textured (fine-grained) stone but with a very uneven surface. The southern transept has two outer orthostats of smooth grey sandstone; the inner pair and end-stone are of smooth reddish sandstone with small quartz inclusions and slight surface knobs and ripples.

Colour contrasts are also visible in the large end-chamber of the southern passage grave, although this has lost its capstones and is open to the elements,

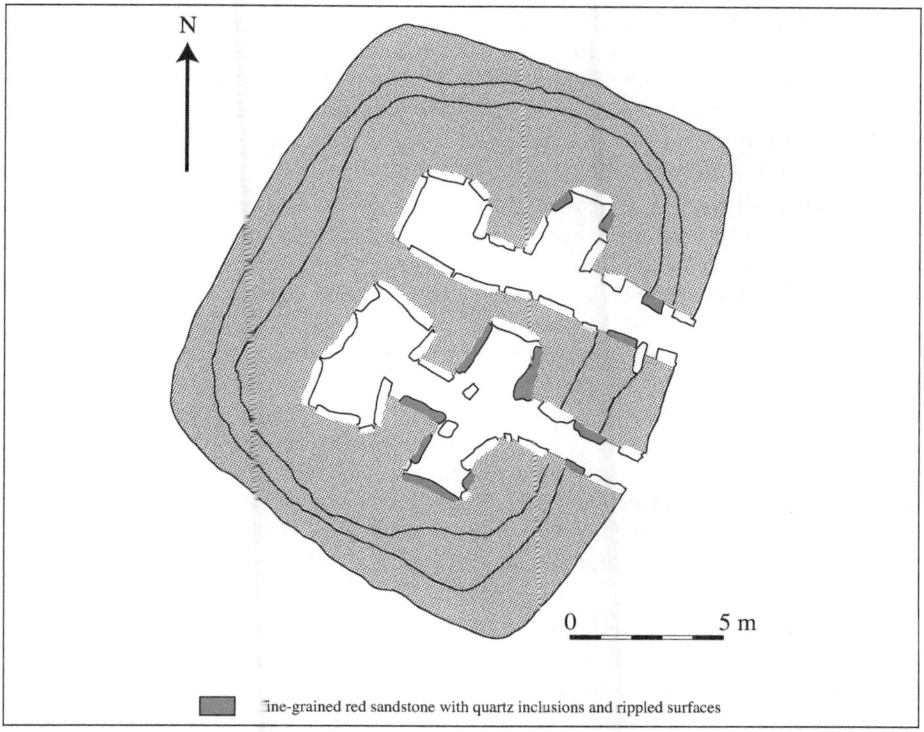

Figure 10.5: Les Mousseaux: distribution of orthostats of fine-grained red sandstone with quartz inclusions.

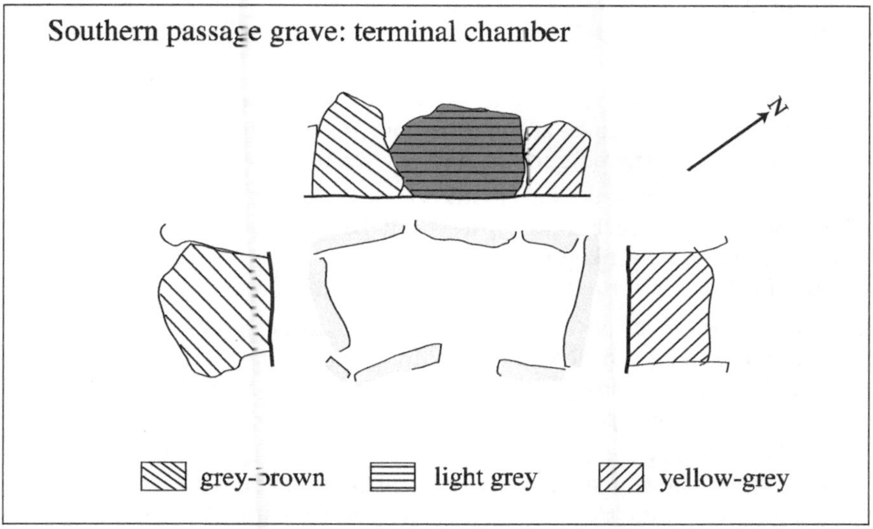

Figure 10.6: Les Mousseaux: stone colours in the terminal chamber of the southern passage grave.

and the stones are accordingly more heavily weathered (**Figure 10.6**). Facing the passage, a large, broad, light-grey slab forms the central visual focus. It is flanked on the left by two slightly knobbly brownish slabs, and on the right by two somewhat smooth yellow slabs. This western wall hence displays the sequence brown-grey-yellow, a sequence that is extended to north and south by a further single yellow and brown slab respectively.

It is such regularities that provide support for the contention that the stones were deliberately arranged according to their colour and/or texture. There remain, however, other issues to resolve. The most important is to establish what it was that led the prehistoric builders to select particular stone sources. In the entrance passages of Les Mousseaux and Les Trois Squelettes, the modern observer sees a pairing and alternation of colours: notably reds, greys and whites. As has been made clear, however, the stones also exhibit significant textural patterning, and it may be that this was as important as the colour patterning, or even more important. For the prehistoric builders, indeed, the significance of these stones may have lain in neither their colours nor their textures but in their places of origin. The sources cannot be precisely identified, but it is clear that they were taken from a number of separate locations within the surrounding area, none necessarily at a distance of more than a few kilometres:

- The white or yellow sandstone could have come from inland exposures near Chéméré, or (more probably) from the cliffs and outcrops of the nearby coast (**Figure 10.7**). Pornic stands on a syncline with beds of white sandstone on its flanks, at La Boutinardière to the east and at Le Redois and La Roussellerie to the north (Ters 1997). Similar limestones are found at the Plage des Moutiers beyond La Boutinardière (Ters 1979). Coastal exposures (in the form of cliffs and rocky outcrops) would have provided the most convenient source of sandstone slabs, and detached blocks are scattered at various points on the beach.
- The red sandstone is less common. It was a formation that developed where sand collected in hollows and depressions on the surface of the micaschist; erosion has left scattered pockets of the material throughout the area.
- The quartzite deployed to such striking effect in the passage graves of Les Trois Squelettes, which is also present at Les Mousseaux (though not used for orthostats), may likewise have come from the immediate vicinity since veins of the material run through the crystalline bedrock.

Thus, the diversity of materials used to build these passage graves makes reference to a diversity of locations within the local landscape, and it may be asked whether it was these locations, rather than the colours and textures in themselves, that were of primary significance to the prehistoric builders.

The distinction is perhaps a little arbitrary: the colour and texture of any particular stone slab must necessarily have acted as cues to its identity and therefore served to indicate the source from which it came. Yet there is an issue of intentionality here that deserves further discussion. Put simply, can we determine whether the stones were selected and placed in position within these monuments in accordance with an established symbolism of colour, as has been suggested in the case of other European megalithic monuments? Might it have been the qualities of the different surfaces and textures that were of special significance? Or was the primary goal to incorporate and combine within these monuments materials that referenced places within the surrounding landscape: places that were perhaps associated with particular social groups or were invested with special mythological significance?

Geology and intentionality

Two examples may help to illustrate this distinction. Some 200 km south-east of Pornic, near Poitiers, stands the Bougon cemetery of megalithic tombs. This group of five chambered mounds was first excavated in the 1840s and again more recently from 1968 to 1986. In the course of the recent excavations the petrology of the orthostats and capstones which make up these chambers was determined, and related to the local geology. This geology comprises a series of limestones: Oxfordian to the east of Bougon, Callovian to the west and south,

Figure 10.7 Location of Les Mousseaux, showing possible sources of white and yellow sandstone. Shaded area indicates intertidal zone.

Bathonian with flints still further west. The precise sources from which the individual slabs had been taken could not be determined, although at a cliff close to Exoudun, one large (3 m x 2 m) slab of Bathonian with flints appeared to have been abandoned in the course of its extraction.

None of the stone used in the Bougon chambers came from a distance of more than 4 km, and sources for all the slabs would have been available within the natural basin formed by the valley of the River Sèvre-Niortaise to the west and south, and the River Pamproux to the north. It is possible that the diversity of materials used in the Bougon tombs reflected the contribution of the different communities living in the area it served. According to this model, the sources of material reveal the catchment area of the communities who built and buried their dead in the tombs. In 1979 this concept was given muscular expression when volunteers from local villages were assembled in order to drag a 32-tonne replica of one of the largest slabs (the capstone of Tumulus F2) along wooden rails (Mohen and Scarre 2002).

The stone slabs in the burial chambers at Bougon display a variety of textures and colours that relate to their geology. Callovian limestone was the preferred material for the chamber orthostats, which in many cases had been shaped and smoothed. In the chamber of Tumulus C, for example, the carefully dressed northern and eastern sidewalls of dark blue-grey Lower Callovian limestone contrast strikingly with the rough-surfaced underside of the yellow-brown capstone of Oxfordian limestone ('aigrain') (Cariou in Mohen and Scarre 2002: 162). This is partly the consequence of intentional smoothing and shaping of the sidewalls, but also reflects geological differences in the quality, colour and texture of the materials. The social catchment model, however, draws attention away from the visual and tactile qualities of the stones and their surfaces and instead emphasises the geography of their sources. It is clear that they came from different sectors of the Bougon landscape, and their combination in these monuments represents the bringing together of material from different places; perhaps places representing different social groups, or with different mythological associations. Above all, however, it was the mechanical properties of the stone which were important: the coarse but strong Oxfordian selected for the capstones, the more brittle but more easily worked Callovian for the uprights. The colours and textures in themselves may not have been the main consideration.

The second example is Stonehenge. The largest stones here are of a fine-grained sandstone known as sarsen, which is probably from the Marlborough Downs some 30 km north of the site. The smaller stones or 'bluestones' have a more varied petrology and a much more distant source in the Preseli Hills of south-west Wales, 240 km from Stonehenge. As Atkinson tells us, the bluestones 'are so called from their colour, which in dry weather is a bluish-grey. But when they are wet after rain they acquire a noticeably blue tinge, especially where the constant abrasion of the feet and hands of visitors has smoothed and even polished their surfaces' (Atkinson 1979: 34). The 'bluestones' do not represent a single type of stone of a particular colour, however, but instead are a collection of stone types of varying surface appearance. Bradley has recently observed how the arrangement of the different varieties of bluestone in the final settings at the site reflects the geography of their origins. The spotted dolerite was gathered together to create the inner horseshoe setting, while the outer ring contained the entire range of non-local stone types (Bradley 2000; Cleal *et al* 1995: 26–29; note that the inner horseshoe also contains a block of unspotted dolerite). Bradley has argued that this arrangement reproduces the distribution of the source rocks in the Preseli Hills: 'The final distribution of the bluestones in the monument at Stonehenge encapsulated the distribution of these different rocks *across their original source area,* with the pieces of spotted dolerite towards its centre and a more scattered distribution of other rock types. In that sense it recreated the distinctive landscape found in south-west Wales in the centre of another, entirely artificial world' (Bradley 2000: 95).

The removal of the bluestones to Stonehenge (accepting that they were transported by human agency rather than glacial action: Green 1997) is hence unlikely to have been undertaken as a constructive act of colour symbolism on Salisbury Plain. We may accept that the colour and texture of the stones are inseparable from their geographical and geological origin; all of these features go together to give the individual bluestones their identity and significance. In this instance, however, the primary importance of the stones appears to have been their exotic place of origin rather than the symbolic significance of their colouration. Stonehenge is certainly not alone in this respect: a similar case can be made for the stone types deployed in the passage graves of Jersey and Guernsey (Bukach 2003).

COLOURS OR CONTEXTS?

We return via these alternative case studies to the fundamental question posed at the outset. Was it the colours of the stones that were important, or other features? The observations discussed in the previous section call into question the idea that the patterning of stone arrangements within the passage graves of Les Mousseaux relates to their colour alone. While early French scholars remarked on colour patterning as a notable feature of the Pornic monuments, and more recent papers by British archaeologists have speculated on the importance of colour symbolism in monument construction in prehistoric Europe (Jones 1999; Trevarthen 2000), it is clear that in at least some cases, stone colour was not in itself the primary consideration, despite the existence of apparent colour patterning. The deployment of red stones in the passage graves at Les Mousseaux appears at first sight to be based on their colour, but the striking textures, ripples and erosion hollows may all have played a part in their selection and inclusion. Furthermore, their mechanical properties – in the way that they split and weather, or their ability to support certain weights or spans – may also have played a part. Relationships between the megalithic slabs and the colours and textures of the contents of the tombs – pottery, chipped stone, human bone (cf Jones 1999) – must also be considered.

Thus it is unwise to conclude too quickly that the colours of the stones used in the Pornic megaliths were of symbolic importance, encapsulating specific values or meanings. To do so would be to privilege vision over the other senses, and to interpret colour difference too readily in terms of colour symbolism. As we have seen, the importance of the Stonehenge bluestones is unlikely to have lain in their colour but may have related instead to the sacred significance either of the distinctive Preseli Mountains, or of some earlier monument in south-west Wales to which they originally belonged. Atkinson remarked that at 1,960 ft, Preseli was the highest point anywhere in south-west Wales, and referred to the source of the bluestones as 'the cloud-capped crest of the sacred mountain far to the west' (Atkinson 1979: 176). It was place rather than colour that was significant. The same may be true of the quartzite that is used to such striking effect at the entrance to the Caveau de la Croix. In a number of ethnographically-recorded societies, quartz and quartzite have been endowed with special power and significance owing to their striking brightness of colour, and perhaps also to the piezoelectrical qualities of quartz rocks which can generate a glow called triboluminescence when they are struck or rubbed together (see Saunders, this volume). In Australia, powerful Ancestral Beings are associated not only with small pieces of quartzite, but also with large, unusual outcrops of sandstone and quartzite that dominate the landscape in western Arnhem Land (Taçon 1991: 196). A systematic association between quartz and shamanism has been noted in western North America (Whitley *et al* 1999).

Seeking to interpret the juxtapositioning of these materials in terms of colour symbolism may hence be to miss the point entirely. These slabs had multiple characteristics, any or all of which may have been of significance to prehistoric societies of 5th and 4th millennium BC Western Europe. The observation that many of the megalithic slabs were taken from their sources and used with only minimal modification (shaping, smoothing or trimming) suggests that the stones themselves in their natural state held special significance (Scarre forthcoming). To human groups with detailed knowledge of their local landscapes, the colour, texture and crystalline inclusions of these slabs will all have served to indicate place of origin. Thus visual and tactile qualities of orthostats and capstones will have created a direct link between the monuments and familiar geological exposures. This will have been particularly the case since the megalithic elements were not quarried from a depth but were taken from surface deposits, cliffs and outcrops. The special qualities or associations of those places are hard to reconstruct: some may have been connected with creation stories or mythological events, others with ancestors or living individuals. Whatever their precise meanings, they probably served to signify identity. Adding to these pre-existing beliefs, the bringing together of stones to build tombs will have generated its own memories and associations, and we may well imagine individual stones being linked with specific groups or individuals in subsequent oral tradition. The patterning of stones within these monuments may indeed have served as a kind of social and ritual mnemonic, the visual and tactile qualities of the stones connecting people with place, and that in turn with ideas, beliefs and memories.

Note

Translations of quotations from French sources are by the author.

References

Atkinson, JC (1979) *Stonehenge*, Harmondsworth: Penguin

Bradley, R (2000) *An Archaeology of Natural Places*, London: Routledge

Bradley, R, Phillips, T, Richards, C and Webb, M (2001) 'Decorating the houses of the dead: incised and pecked motifs in Orkney chambered tombs' 11 *Cambridge Archaeological Journal* 45–67

Bukach, D (2003) 'Exploring identity and place: an analysis of the provenance of passage grave stones on Guernsey and Jersey in the Middle Neolithic' 22 *Oxford Journal of Archaeology* 23–33

Cassen, S, Defaix, J and Mens, E (2001), 'Pornic, Les Trois Squelettes', in *Bilan Scientifique de la Région Pays de la Loire 1999*, Nantes: Direction Régionale des Affaires Culturelles/Service Régional de l'Archéologie 17–18

Cleal, RMJ, Walker, KE and Montague, R (1995), *Stonehenge in its Landscape*, London: English Heritage

Cummings, V (2002) 'Experiencing texture and transformation in the British Neolithic' 21 *Oxford Journal of Archaeology* 249–61

Daniel, GE (1939) The transepted gallery graves of western France' 5 *Proceedings of the Prehistoric Society* 143–65

De Wismes (1876) 'Le tumulus des Trois Squelettes à Pornic (Loire–Inférieure)' 15 *Bulletin de la Société Archéologique de Nantes* 199–271

Devignes, M (1996) 'Les rapports entre peintures et gravures dans l'art mégalithique ibérique', in *Art et Symboles du Mégalithisme Européen*, RAO supplément 8, 9–22

Drewal, HJ and Mason, J (1998) 'Beads, body and soul. Art and light in the Yoruba universe' *African Arts* 18–27

Green, CP (1997) 'The provenance of rocks used in the construction of Stonehenge', in Cunliffe, B and Renfrew, C (eds), *Science and Stonehenge*, Oxford: OUP for the British Academy, pp 257–70

Jones, A (1999) 'Local colour: megalithic architecture and colour symbolism in Neolithic Arran' 18 *Oxford Journal of Archaeology* 339–50

L'Helgouach, J (1977) 'Le cairn des Mousseaux à Pornic (Loire-Atlantique). Nouvelles fouilles et restaurations' 104 *Bulletin de la Société Polymathique du Morbihan* 161–72

L'Helgouach, J (1979) 'Circonscription des pays de la Loire: Loire-Atlantique' 22 *Gallia Préhistoire* 557–62

L'Helgouach, J, Le Gouestre, D and Poulain, H (1989) 'Le monument mégalithique transepté de la Joselière (ou du Pissot), au Clion-sur-Mer (Loire-Atlantique)' 6 *Revue Archéologique de l'Ouest* 31–50

L'Helgouach, J and Poulain, H (1984) 'Le cairn des Mousseaux à Pornic et les tombes mégalithiques transeptées de l'estuaire de la Loire' 1 *Revue Archéologique de l'Ouest* 15–32

Lisle du Dréneuc, P de (1882) *Dictionnaire Archéologique de la Loire-Inférieure (époques primitive, celtique, gauloise et gallo-romaine)*, Nantes

Lisle du Dréneuc, P de (1892) 'Notice sur les fouilles du tumulus de la Motte, Sainte-Marie (Loire-Inférieure)' 32 *Bulletin de la Société Archéologique de Nantes et du Département de la Loire-Inférieure* 199–203

Mohen, J-P and Scarre, C (2002), *Les Tumulus de Bougon. Complexe mégalithique du Ve au IIIe millénaire*, Paris: Errance

Scarre, C (2002a) 'A place of special meaning: interpreting prehistoric monuments through landscape', in David, B and Wilson, M (eds), *Inscribed Landscapes: Marking and Making Place*, Honolulu: University of Hawaii Press, pp 154–75

Scarre, C (2002b) 'Epilogue: colour and materiality in prehistoric society', in Jones, A and Macgregor, G (eds), *Colouring the Past*, Oxford: Berg, pp 227–42

Scarre, C (forthcoming) 'Displaying the stones: the materiality of "megalithic" monuments', in DeMarrais, E, Gosden, C and Renfrew, C (eds), *Rethinking Materiality: The Engagement of Mind with the Material World*, Cambridge: McDonald Institute for Archaeological Research

Shee-Twohig, E (1981) *The Megalithic Art of Western Europe*, Oxford: Clarendon

Taçon, PSC (1991) 'The power of stone: symbolic aspects of stone use and tool development in western Arnhem Land, Australia' 65 *Antiquity* 192–207

Ters, M (1979) *Carte Géologique de la France, XI-24. Machecoul. Notice explicative*, Orléans: Bureau de Recherches Géologiques et Minières

Ters, M (1997) 'Itinéraire 10. La côte du pays de Retz', in Gabilly, J and Cariou, E (eds), *Poitou, Vendée, Charentes. Guides Géologiques Régionaux*, Paris: Masson, pp 142–49

Tessier, M (1994) *Dictionnaire Archéologique du Pays de Retz*, Nantes: Société Nantaise de Préhistoire

Tilley, C (1996) *An Ethnography of the Neolithic. Early Prehistoric Societies in Southern Scandinavia*, Cambridge: CUP

Trevarthen, D (2000) 'Illuminating the monuments: observation and speculation on the structure and function of the cairns at Balnuaran of Clava' 10 *Cambridge Archaeological Journal* 295–315

Turner, V (1967) *The Forest of Symbols. Aspects of Ndembu Ritual*, Ithaca: Cornell University Press

Whitley, DS, Dorn, RI, Simon, JM, Rechtman, R and Whitley, TK (1999) 'Sally's Rockshelter and the archaeology of the vision quest' 9 *Cambridge Archaeological Journal* 221–47

– 11 –

READING THE EARTH

PHILOSOPHY IN/OF THE FIELD

Robert Frodeman

Our view of science has been dominated by a prejudice: the philosophy of science has been taken to mean the philosophy of *lab* science. Field-oriented sciences, insofar as they have been considered at all, have been viewed as merely an inferior form of what properly happens in the laboratory. This is an epistemological error rife with social and political consequences, for the mineral world is best approached *in situ*.

Rather than the laboratory, the native home of the philosophy of science lies in the field. In its best dress, the philosopher wears hiking boots and carries a walking stick. The terrains to be covered by a philosophy of the field are quite varied – field biology and archaeology, botany and soil science – and involve the cultivation of the eye and hand and limbs in concert with the measuring tape and proscenium. This essay explores one such terrain, field geology, in order to consider how it is that people come to know and exploit the mineral world in particular ways.

THINKING ABOUT THE EARTH

Twenty five hundred years ago the Greeks began a search for *logos* – a rational pattern to the world. By the time of Plato, Greek thinking focused upon a specific type of *logos*. Turning our attention away from our bodies, and from the body of the earth, the rational gaze was directed upward, toward the stars, and inward, toward a pure mental order. Thus the Greek search for *logos* became a quest for a very particular type of order – one that was distant, regular, immutable, and certain.

The motivation behind this is clear. As Plato notes, our knowledge of things in the sensible world is always questionable. How can we claim to truly know something that is constantly changing? Temporality was seen as the enemy of rationality, rendering every claim to truth inconclusive and suspect. Knowledge must be grounded in those realms beyond the corrosive effects of temporality. Only in the realm of thought – and in the celestial sphere, a region considered beyond material corruption – could we find the conditions suitable for truth.

The creation of mathematics and the identification of celestial order with reason were watershed events in the unfolding of human rationality. By defining reality as the rational, and identifying rationality with the regular, immutable and certain, the reality of the sensuous world was denigrated or even denied. Through its long history Western culture has sided with this view. As inheritors of this tradition, we have valued the enduring over the transient and the celestial over the earthly, down to the present-day priority our culture gives to mathematics and the mathematics-based sciences. Quantity and objectivity – the former understood as the deep essence of things, and the latter defining knowledge in terms of those things that we can turn into objects – remain the tokens of the real.

Our embrace of this notion of rationality has placed us in a peculiar situation. On the one hand, it has given us an unprecedented control over the world, resulting in the massive amelioration of physical suffering. But this success has come at a price. This view of rationality has deformed our personal and social relations, as questions of ethics, politics, metaphysics, aesthetics and religion – those areas that cannot be parameterised and distanced, and which are immersed in time and contingency – are viewed as marginal and even nonsensical. Moreover, this conception of rationality, as it has pursued a perfect understanding and perfect control of the world, has severely compromised the ecological order upon which all life depends. The romanticism that long coloured our view of non-Western and pre-technological cultures – seeing them as innocent and benign stewards of the land – has largely dissipated. Nevertheless, it remains clear that our scientific and technological culture has inaugurated a dangerous new epoch in our relation to the earth.

It is noteworthy that we have no single discipline that seeks a unified understanding of our relationship with the earth. We might suppose that geology would be this discipline. Taken at its word, the field would provide a complete *logos* of the planet the Greeks called Gaia: synthesising geoscientific research with poetry, nature writing and hermeneutic analysis, and combining these with geopolitical considerations of issues such as resource depletion, pollution and climate change. But until quite recently our epistemological presumptions have encouraged geology to stay within what count as normal boundaries. The radical interdisciplinary potentialities of the field have gone unnoticed. In fact, geology – and the field sciences generally – has been ignored by philosophers of science, who have instead focused on physics as the paradigmatic science.

Our Platonic prejudices have discouraged us from inquiring too deeply about the earth. While the study of the heavens has been pursued for millennia, the systematic examination of the earth is barely 200 years old, beginning with James Hutton in the 1780s. This matched our prejudices: in contrast to the clarity of mathematics, and the transparency of the mathematicised heavens, the earth was impenetrable, and subject to sudden violent alterations. The earth was 'beneath' us – a subject too 'earthy' and unstable to be worthy of serious attention.

Not coincidentally, geology has received little attention from the humanities.[1] Certainly contemporary philosophy has not recognised the field as a subject worthy of reflection. There is no philosophy of geology or of the earth sciences as there are philosophies of physics and of biology. The two main schools of contemporary philosophy, Analytic and Continental, have ignored geology. It has been assumed – few even thought to argue the point – that examining the earth sciences was unnecessary for understanding the nature of either geology or science in general. The few statements that one can find by philosophers on the subject of geology sound a common refrain: 'Geology is a science just like other sciences, for example physics or chemistry' (Watson 1969).

This de facto reduction of geology to physics – seeing field science as a poor reflection of what properly happens in the lab – begs a large set of questions concerning the proper compass of the philosophy of science. Why, after all, do we view the philosophy of science as primarily a species of epistemology, rather than of political philosophy, metaphysics, or the philosophy of religion? For the social and political consequences of science, as well as its crypto-metaphysical and theological aspects, are at least as salient as the epistemological ones. A thorough review of this question cannot be encompassed in the brief span of this essay. But a partial answer may be found in considering the role of paradigmatic examples for our understanding of the nature of science. Relying upon examples taken from the heavens and the laboratory, rather than out in the field, philosophers of science have facilitated a closeted image of the nature of science that turns away from our lived experience of the mineral world.

PHILOSOPHY AND ITS ACCOUNT OF SCIENTIFIC REASONING

Like every discipline, philosophy has its presuppositions. Traditionally, one of these presumptions is that philosophy properly begins in abstract theorising distanced from personal needs, the hurly-burly of the marketplace, and the claims of politics. This distance (nowadays described as an unrealisable ideal, but worthy of pursuit nonetheless) is supposed to deliver philosophical reflection from idiosyncratic personal influences, while ensuring the greatest degree of objectivity and universality. It is a stance that has discouraged philosophers from paying close attention to the examples they rely upon to illustrate their theories.

Much of Derrida's early work turns on this point, where he explores the ways in which philosophy's central concepts have often consisted of blanched or 'white' mythologies where old metaphors gradually become effaced through use (Derrida 1982). From an entirely different perspective, a wide body of recent psychological and cognitive research has also supported this account, arguing that thinking (scientific or otherwise) is fundamentally symbolic and analogical in nature (Lakoff and Johnson 1999; Minsky 1988). Humans think in images: concepts are created through an analogical process, as we generalise from an example, having it stand as an image or symbol of a new case or of a larger truth (Vosniadou and Ortony 1989). There is nothing wrong with, or even particularly limiting about, our reliance upon images to understand the nature of things. It is simply the nature of human cognition. But it does highlight the fact that we should be aware of the examples we rely upon.

Philosophers of science have sought to provide a formal account of the nature of scientific reasoning. But in doing so, they have also illustrated their theories with a few notable examples taken from the history of science. These examples have overwhelmingly centred upon two places, the heavens and the laboratory. Consider first the heavens. It is a commonplace of intellectual history that Galileo and Newton shattered the medieval conception of the universe. Before their work, the cosmos was thought to consist of two regions: a sublunary realm subject to change and imperfection, and a celestial realm that was incorruptible and the home of the gods. Galileo and Newton demonstrated that terrestrial and celestial space were in fact homogeneous. The same laws of gravity governed both realms, and both were subject to the effects of time. The discovery that terrestrial and outer space were uniform meant that every position in the universe could be measured and calculated. A distinct realm of the divine was banished, as the laws of physics now situated everything.

But while in one sense Galileo and Newton destroyed the distinction between the terrestrial and celestial realms, in another sense they described a space for science that preserved many of the traditional features of the heavens. For the laws of physics abstract from the unmanageable, experiential and deeply historical reality that is our common destiny. In its logical clarity, and in the view of science as distanced from our personal and political lives, the stories of Galileo and Newton offer an image of science with many of the characteristics of medieval celestial space. After all, cannonballs of different weights *do* fall at different rates from the tower at Pisa: it is only by placing them in an idealised space (ie, one without friction) that Galileo could identify a law that describes how they always hit the ground at the same moment. Similarly, Newton's laws of celestial mechanics offer a powerful image of scientific reasoning, but they also provide a convenient example of scientific certainty that is distanced from everyday affairs. Science viewed from the perspective of Newton's Laws is disconnected from the traffic of everyday life, formulated in a rigorously mathematical way, and summed up in a few equations. Newton's research is an admirable example of the power of a certain kind of science. But how appropriate is the model, given the types of challenges science must address today?

This is not to suggest that the only reason that Newton and Galileo are used as (literally) textbook examples of scientific advance is because their work fosters a particular image of science as both certain and disassociated from everyday life. But we must ask why these theories have the place they do in our accounts of science. Why, for instance, does nearly every college course in physics begin with Newtonian mechanics, rather than with an account of quantum mechanics, electricity or optics? Newton's work is our paradigm of scientific knowledge: powerful, mechanistic and certain (once we set aside the real world variables), and existing in an ideal, mathematical space. Newtonian mechanics thus fulfils both science's and society's dream of a scientific truth that does not entangle us in the complications of our social lives.

Of course, science did not stay in the heavens. But when the work of science was brought down to earth, it was located in a space that retained many of the characteristics of celestial space: the laboratory.

The laboratory is, by definition, an unreal space. Within it, conditions are parameterised, objects are constrained, and materials are purified. The space of the lab is utopian: in Greek, *ou-topos* means literally a 'no place'. The lab is a space abstracted from the rest of the world; it makes no difference whether a lab experiment is conducted in St Louis or Moscow. And not only is the space of the lab idealised; in the laboratory, time is idealised as well. In the lab, history doesn't exist. Time exists, of course; it takes time to run an experiment. But the laboratory eliminates history – the non-repeatable and contingent flow of events, both natural and cultural, within which we live our lives. The defining characteristic of laboratory science is that the conditions and results can be repeated on demand; and thus the same zero-point, purified of antecedents, the same erasure of the past always remains as an option.

The situation within the lab may be summarised as follows: what we gain in control, we lose in realism. The irony here is that while the lab sciences are considered the gold standard for scientific truth, the findings of the lab often don't apply realistically to the social controversies we face. The question that should lie at the centre of the philosophy of (lab) science, but is too often treated as incidental, is how do we relate the closed, ahistorical system of the lab to the open and historical world that we live in?

This question has received scant attention from philosophers of science. And no wonder. For attention to this issue would force the philosophy of science to recognise how much the lab sciences depend upon analogical reasoning. Judging how the results from the lab's closed and controlled system relate to the open and historical world we inhabit demands skill at drawing apt analogies. Now, a great deal of work has been done on the role of analogy within scientific reasoning (eg Hesse 1966). This work, however, has been concerned with the cognitive or psychological role of analogy in the creation of concepts – how analogies help us imagine scientific problems – rather than with the analogical relation between lab and reality. And while the exact relation between lab and field is a vexing question, this much is clear: it will not be susceptible to mathematical formulisation. Rather than mathematics, analogical reasoning finds its home in disciplines such as philosophy, history, aesthetics and semiotics. Indeed, while the strength of analogies is often measured by counting the number of elements shared in common, the power of an analogy can also turn on the interplay between the one or few relevant elements existing in dynamic tension with a larger set of dis-analogous elements.

Our reluctance to think through the lab sciences' fundamental dependence upon analogy with the field has given birth to two kinds of mischief. First, when society fails in its attempt to adequately apply laboratory results to the world in which we live, it encourages disillusionment with science. And second, this reluctance sponsors the devaluation of other types of knowledge – in particular, the historical sciences and the non-quantifiable humanities – when they, by their very nature, fail to live up to the standards of the lab. The result of centring the philosophy of science on the two spaces of the heavens and the lab has been the creation of an enormously powerful intellectual apparatus with a very tenuous relation to social concerns. In this way, the philosophy of science has come to mirror the examples it has chosen.

SCIENCE IN THE FIELD

Basing our image of science in the heavens and the lab has created a metaphorics of science as cerebral, ahistorical, and isolated from the other aspects of our lives. In contrast, a field-based philosophy of science offers us a more socially engaged and epistemologically realistic image of science. To see why, consider the distinctive nature of a field science such as geology.

Time spent with a field geologist proceeds at a different rhythm. Outside, in the open air, we are exposed to all the natural elements: sun and sky, wind and rain, and a natural symphony of sound. Rather than the controlled experience of the lab, the field scientist is immersed in a constantly changing sensorium that promotes and sustains the wonder and serendipity of experience. The walls – conceptual and physical – that isolated the lab scientist are now gone, as the field scientist passes through a shifting, frameless flow of events. In the lab, nature is constrained so that it will show itself in a regular fashion. In the field, the scientist must adjust to nature's patterns, and cultivate a disposition of alert repose, anticipating the moment when the world reveals itself. The biologist may have to wait all day for a bear's chance appearance, and the geologist may need to devote hours splitting bedding planes in a shale unit, searching for the telling fossil.

While the lab narrows and controls the flow of information, the field's perceptual and conceptual kaleidoscope exceeds our capacities to self-consciously sort, test and categorise it. Field scientists must thus develop intuitive skills for parsing knowledge in implicit, non-propositional ways. Walk an outcrop with a seasoned field geologist. Years in the field have taught him how to sift through the superabundance of information and identify the significant anomaly – an odd dip in the strata, an unusual colour along the bottom of a layer.

Nor do field scientists operate in the objective geographical space conceptualised by Cartesian geometry. Rather, they move through a world of implied meaning that is constituted by the acting and perceiving body as it interacts with its environment. In geological fieldwork, the paired limestone and shale beds that are easily distinguishable at 50 ft grow indistinct at 6 ft, only to resolve themselves again at 12 inches. Similarly, the land unfolds as one passes over it, revealing an infinite number of perspectives that the geologist's mind integrates and contrasts. Of course, out of this the geologist must abstract a testable hypothesis and numerical values. But these results are built upon the intuitive use of judgment in which the geologist selects – and constructs – a system of signs and blends multiple perspectives from a nearly infinite amount of potential data.

The field scientist's interaction with the earth depends upon what Merleau-Ponty (1962) called 'motor-intentionality'. Motor-intentionality refers to the directed state of action/perception that operates before we consciously posit an object for examination. In driving a car, for example, we do not consciously analyse traffic patterns and driving conditions. The experienced driver does not stare at the road, but rather scans the field of vision in search of clues: the faded yellow and white patches of the street lines, the paired flashes of animal eyes, the flicker of tail lights, the telltale glimmer of bike metal. Like a surfer or a musician, the driver is not so much following a set of logical procedures as moving along in a flowing river of experience, on the lookout for clues that point toward alternative futures. Indeed, while driving, it is dangerous to focus upon a single object, thereby losing the ambient awareness of one's surroundings. Practiced drivers do not experience a distinct separation between themselves and their automobiles. Such a separation only comes into play on a conceptually and linguistically objectifying level of consciousness. Similarly, fieldwork often operates below the level of fully conscious intentional activity, which involves concepts and introspective reasoning, and above such mere reflex movements as the blinking of the eyes or the heartbeat (Raab and Frodeman 2002).

Of course, this intuitive awareness of the environment is itself the result of extensive training. Dreyfus and Dreyfus (1986) offer a phenomenological account of the stages in skill acquisition from novice to expert. In learning to drive a car, the novice starts with conscious, rule-governed behaviour: 'allow one car length between you and the next car for every ten miles per hour.' Gradually such rules fade, replaced by immediate, intuitive responses to situations perceived in a holistic manner. In terms of our lived experience, the car becomes more and more an extension of our body rather than an object external to it. Field scientists – indeed, all scientists, and everyone who masters a craft – move through a similar set of stages. In the beginning, field scientists spend great amounts of time in the field consciously applying rules and guidelines. For the experienced fieldworker, however, the work proceeds smoothly and automatically, with little conscious reasoning.

The characteristics of science as it is practiced in the field provide us with an image of scientific reasoning that is much closer to the methods of everyday reasoning at work in our personal and political lives. Field experience presents inescapable spatial and temporal ambiguities. Geologists cannot subject their object of study to the controls of the lab; nor can they duplicate the conditions they find a potentially infinite number of times. Rather, field scientists engage in a deeply human activity of interpretation, following out the logic of signs in which they ascertain relationships that are implicit rather than overt.

It is for precisely these reasons that field science has been neglected by the philosophy of science. The paucity and ambiguity of the data, and the non-repeatability of the results, have been considered a weakness compared with the standards achievable by laboratory experiments. But this reaction should be stood on its head. By mastering such skills as working with an overabundance of ambiguous evidence within an essentially historical milieu, field scientists powerfully model the complexities of reasoning in science and in our personal and political lives. We can understand the basic meaning of a term, or adequately manipulate a set of symbols, without having a rich, holistic sense of how a point properly plays out in practice. We can know the rules of a game without being good at playing it, or know basic principles of how to read rocks without being able to properly apply them to a particular outcrop. In the midst of a game, an athlete cannot explicitly conceptualise the flow of information coming in from all directions. The situation is too complex, and the data received are obscured (for example by the roar of the crowd, or a pair of hands in the face). But amidst this confusion, the athlete is still able to perfectly place a pass.

The contrasts drawn here can be overstated; all of the skills identified with field sciences such as geology appear to one degree or another in the laboratory as well. But they appear most vividly within fieldwork. It is also worth emphasising that the distinction between lab and field science runs *through* as well as *between* the various sciences. Archaeology and geology both have significant laboratory components, while there is a field element to atmospheric chemistry. Since the subject of every scientific discipline is nature, even those fields most immersed in laboratory life must connect at some point with field experience. But we can still usefully distinguish between those disciplines that are more laboratory-based (eg physics, chemistry) and those whose work is much more closely tied to experience in the field (eg geology, ecology).

In his autobiography, Charles Darwin recalls going into the field in Wales in 1831 with Adam Segwick, nearly a decade before Louis Agassiz's theory of continental glaciation:

> Neither of us saw a trace of the wonderful glacial phenomena all around us; we did not notice the plainly scored rocks, the perched boulders, the lateral and terminal moraines. Yet, these phenomena are so conspicuous that ... a house burned down by fire did not tell its story more plainly than did this valley. (Stoddart 1994)

But while this experience is ubiquitous among geologists, the process by which one moves from blindness to insight has been little analysed by either science or the philosophy of science. Nonetheless, this experience is present in reading a poem and in judging the intonations and body language of a salesman. The act of semiosis – learning to see something as a sign expressing meaning – is arguably humanity's most basic rational activity. It is this skill, rather than propositional or predicate calculus, that should be taught in logic classes; indeed, it should find a home in every discipline across the curriculum. Skill in semiosis is found in the scientist's work in the lab or the field, the physician's ability to recognise signs of incipient illness, and our ability to read the motivations of the people we meet.

Like vision itself, visual intelligence depends upon a kind of background awareness. Vision is not structured in the subject-verb-object terms of our language. The logic or structure of an image is a figure-background relation, a point of focus surrounded by a penumbra of generalised awareness. Of course, we can shift our point of focus to a place that previously was background. But we cannot focus upon the background as such. This ambient awareness of one's surroundings involves another, more subtle and intuitive type of *logos* – an orienteering type of knowledge where we come to know our way around a subject. Knowing your way around the Grand Canyon – or New York City – does not mean that you are familiar with its every nook and cranny. But it does imply that you are oriented in that space, and that if you were dropped into an unfamiliar side canyon, you would know where to look for water or where to find a break in the cliffs to climb out.

Visual intelligence involves more than the use of our eyes. Spatial understanding is kinetic; to understand three-dimensional space, one must move through it. Geology involves a tactile and kinesthetic type of reasoning, demanding an active and mobile body rather than the disembodied intelligence of mathematical knowledge. One must walk – and climb – the cliffs in order to understand them, to see them from a variety of angles and perspectives. The geologist stares at the outcrop, of course. But he also moves around in order to see it at different angles. He strikes or digs the rock out, feeling its density and resistance. He directs his body back and forth, left and right to take in the different geological features, intuitively seeking an optimum distance and observational angle for the features being considered. When we view an object, we seek 'a direction viewed from which it vouchsafes most of itself'. Merleau-Ponty (1963) calls this the tendency toward gaining a 'maximum grip' upon the environment. In some cases, however, there is no one maximum grip. We must try to hold on to a series of irreconcilable views of an object.

To grasp the relation between the visible and the invisible, consider the work of the 19th century French painter Paul Cézanne, especially his repeated rendering of Mont Sainte-Victoire near Aix-en-Provence (see **Figure 11.1 colour section**). Cézanne dismissed the title of 'Impressionist'. He described his goal as painting how we actually see the world, presenting the lived perception of an object rather than a geometric or photographic one. He remarked to his friend Bernard, 'we must develop an optics, by which I mean a logical vision – that is, one with no element of the absurd' (Johnson 1993). Cézanne's wife described his manner of working: 'He would start by discovering the geological foundations of the landscape, convinced that these abstract relationships, expressed, however, in terms of the visible world, should affect the act of painting' (*ibid*). For Cézanne, the surface of an object is not simply a two-dimensional flat plane, but rather the surface obscuring a hidden depth, which reaches its expression at the surface.

Cézanne's goal was a rigorous fidelity to the object itself. But instead of pursuing this by drawing the outlines of things and then colouring them in, Cézanne constructed objects not by lines but by colour. 'The contour of an object conceived as a line encircling the object belongs not to the world but to geometry ... the world is a mass without gaps, a system of colours across which the receding perspective, the outlines, angles, and curves are inscribed like lines of force ...' (*ibid*). An object's shape is constructed from the inside out; the object's edge is an ideal limit rather than a line, depicted through a series of receding contours. 'Nature', Cézanne urged, 'must be seen from the inside' (*ibid*).

Far from being an impressionist's reverie, Cézanne's phrase is a precise account of the geologist's actual experience and procedure in the field. Rather than simply collecting data, the geologist *envisions* the outcrop, struggling to see it as a system of colours or of shapes, but in any case as a system that evolves into a template or *logos* that the facts will fit within and modify. The geologist eventually converts this initial experience into the languages of mathematics and geometry, making it both precise and verifiable. But the essential act here is the act of envisioning.

In his working notes *The Visible and the Invisible*, Merleau-Ponty (1968) explores how it is that what we do not see guides, relates to, or interacts with what we do see. In vision, there is a relation between the surface of things and the depths hidden to us. In sketching a face, the artist tries to express 'the surface of a depth' – the inner nature of the individual that finds its expression at the surface (*ibid*). In a similar fashion, the geologist gives us 'a cross section of a massive being', interpreting the two-dimensional surface as an expression of a three-dimensional structure. Walking Six Mile Fold in Colorado with a structural geologist, one struggles to see a two-dimensional surface in terms of the underlying relationships. What was objectively there was a sinuous limestone bed outcropping across the surface of a ridge. But once it is seen from the inside, this becomes a fold bisected by a stream. The geologist sees *into* the rock: translating two dimensions into three; grasping the nature of an object or pattern by mentally rotating, unfolding, or completing it in space; relying upon his knowledge of the outcrop's geochemistry to inform an interpretation; or evaluating the present outcrop in terms of others across a region. In all of these cases, the geologist's vision is structured or guided by what is not visible. The practiced geologic eye has learned how to penetrate the object – probing, testing it, sizing it up – seeing the surface as a surface of a depth, rather than simply a surface.

The eye is not merely a receptor of data; vision both encapsulates and penetrates its object. Significant seeing includes a background or ambient awareness as well as focusing upon the object. The geologist asks pertinent questions of the outcrop by knowing his way around a topic, understanding it in an intuitive way. Geologic seeing depends upon the imaginative capacity to envision the mineral world, rather than the straightforward application of scientific principles. Geologic seeing is poetic vision constrained by the sobriety of science, a series of daring imaginative leaps disciplined by examination and measurement. Defying the stereotypes of science, field geologists remain a clan of surreptitious poets. Their original aesthetic and metaphysical delight in landscapes still beats beneath their discipline's technical terms and laboratory apparatus.

THE POLITICS OF GEOLOGIC TIME

The mineral world, of course, lies at the root of all economic activity. This is clear enough when we are considering petroleum, copper or coal; but it's less obvious that corn is geologic in origin. Soils are formed through geologic processes over geologic periods of time: in the High Plains, some of the best soils on the planet consist of loess deposits, the products of Pleistocene glaciers. Loess consists of silt layers, sometimes tens of feet thick, in areas that once bordered continental glaciers. In the Pleistocene, glaciers ground the rock beneath them into tiny particles known as rock flour. The glacier's melting edge then released large amounts of the flour, which was carried by the wind in huge dust storms and deposited in downwind areas. Today 25% of the High Plains' topsoil is seriously degraded, with large amounts having washed down to the Gulf of Mexico in the 125 years since European settlement – a permanent, irreplaceable loss, from the perspective of human interests. Moreover, the groundwater used to irrigate these crops is also geologic in origin, fossil water drawn from the Ogallala aquifer, the legacy of glaciers that melted 18,000 years ago. It is estimated that at current rates of drawdown the Ogallala aquifer will last another 30 to 50 years (Brown 1997).

But if geology has one foot rooted in economics, its other rests in an experiential, poetical space that also goes to the origins of the discipline. The historical geologist's work is unusually dependent upon the capacity to dream. Of course every human endeavour, and certainly every science, requires imagination and creativity. But geology embodies this point in a singular way. It is too often forgotten that, when we look at re-creations of dinosaurs and the environments they inhabited, these rich interpretations are based upon an extended series of inferences drawn from fragments of crushed bone pulled from twisted and fractured strata.

This process of dreaming the outcrop itself depends upon a type of ontological disruption, for the discipline of geology, in its development from the field of mineralogy, displaced our human-scaled sense of time. James Hutton's (1997) discovery of geologic time – a length of time so expansive that he could imagine 'no vestige of a beginning, no prospect of an end' – represents much more than an arithmetic fact. Opening ourselves to the possibility of making sense of an outcrop requires a transformation in one's lived sense of time, and thus of reality. As John McPhee demonstrates in his series of books on geology and geologists, walking the earth and thinking in terms of deep time profoundly affects the way one experiences the world (McPhee 1983). What was taken as real becomes ephemeral, while the inconsequential becomes laden with significance.

Viewed geologically, building condominiums on a beachfront is akin to setting up lodging on an iceberg; the very term 'beachfront property' becomes an oxymoron. And commonplace acts such as tossing away a Styrofoam container that held a burger become absurd and even obscene, discarding after a moment's convenience a part of the world's shrinking petroleum inheritance that came down to us through a 150 million year process.[2]

Geologic time's effect, however, does not turn on the objective fact concerning a particular rock's age. In terms of our lived experience, it makes no difference whether the earth's age was finally found to be 2 billion or 4 billion years old. For in either case, we have utterly transcended our previously anthropocentric view of time. If the earth is something like 6,000 years old, humans fill most of history's canvas. If the world is hundreds of millions or billions of years old, clearly we are a small part of a much greater story. When we see ourselves as geologic phenomena, placed within a grand cosmological and geological narrative, our lives and our behaviour take on a different colour.

Such geologic narratives don't exist only on the grand cosmological scale. They get their greatest grip upon public policy on the local level. Helping people place themselves in terms of both space and time is a powerful means of addressing environmental problems. If you live in Phoenix, understanding the basic hydrologic story of central Arizona – the irrigation efforts of the Hohokam, the drying up the Salt River west of the Mazatzal Mountains, and the huge periods of geologic time that it took to create the decidedly limited aquifer underlying the area – is a powerful first step in getting people and municipalities to change their behaviour. Rather than trying to impress ethical principals upon people, such knowledge changes our experience of reality, and thus our behaviour.

The implications of geologic knowledge are as much poetic as political. For our understanding of geologic time comes to us through the use of metaphors. So, for instance, the earth's history is compared with the 24 hours of a day, where the last five minutes before midnight correspond to the appearance of humanity. Just as we can repackage one million seconds into 11 days, and one billion seconds as 32 years, in order to make these large numbers intuitively graspable, we can order geologic time through terms like Cretaceous and Jurassic to give us an intuitively understandable narrative framework for geologic time. If days and years are dictated by the rhythms of the solar system, geologic periods are defined by crises in the history of life.

The seeming incomprehensibility and irrelevance of geologic time most clearly passes into something meaningful at those points where human and geologic time intersect. Earthquakes, floods, hurricanes and droughts are moments where deep time erupts into our more familiar temporal rhythms. These points of contact will become increasingly common as the pressure of population growth and technological potency make the intersection of human and geologic time a common territory. Geo-economics and geo-philosophy will come together in the place where all questions eventually play out: the communal place that we may call the geopolitical. So, for instance, when irrigation west of the 100th meridian within the US is no longer possible because the Ogallala aquifer has run dry, or development in southern Arizona is tied in knots by water shortages, we will have felt geologic time in the most palpable of ways.

CONCLUSION

The mineral world is not simply something other to us. Rather than Cartesian *res cogitans*, we are embodied beings in constant interaction with and moving through a world largely consisting of stony parts: cliffs and mountains, stream bottom and bedrock, building stone and our own bodies. A philosophy of science worthy of the name must account for these interactions rather than simply retreating to the artificial world of the laboratory.

A mineral is commonly defined as 'a naturally occurring homogeneous solid, inorganically formed, with a definite chemical composition and an ordered atomic arrangement' (Mason *et al* 1968). But the line between self and other, the organic and inorganic, cannot be as easily drawn as that. The vast majority of limestone is organic in origin, consisting of clumps of calcium carbonate ($CaCO_3$) secreted by marine organisms such as corals and foraminifera. Humans also produce hard mineral parts, both during development and throughout life – the latter point familiar to victims of broken bones and kidney stones. Attending to the identities between our selves and the mineral world holds the promise of a more engaged and vigorous environmentalism, less dependent on the predictions of the lab, and more focused on our carnal nature.[3]

Notes

1 Possibly the most humanistic treatment of geology is contained in the writings of John McPhee (see eg McPhee 1981, 1983 and 1993).

2 Styrofoam, properly called polystyrene, is a petroleum by-product that is used in a variety of products including plastic utensils and television sets.

3 This essay has been adapted from *Geo-Logic: Breaking Ground between Philosophy and the Earth Sciences* (Frodeman 2003).

References

Brown, L (1997) *The State of the World*, San Francisco: Worldwatch Institute

Derrida, J (1982) *Margins of Philosophy*, Bass, A (trans), Chicago: University of Chicago Press

Dreyfus, HL and Dreyfus, SE, 1986, *Mind over Machine: The Power of Human Intuition and Expertise in the Era of the Computer*, New York: Free Press

Frodeman, R (2003) *Geo-Logic: Breaking Ground between Philosophy and the Earth Sciences*, Albany, NY: SUNY Press

Hesse, MB (1966) *Models and Analogies in Science*, Notre Dame: University of Notre Dame Press

Hutton, J (1997) [1797], *The Theory of the Earth, with Proofs and Illustrations in Four Parts*, Volume 2, Geikie, A (Sir) (ed), London: Geological Society

Johnson, GA (ed) (1993) *The Merleau-Ponty Aesthetics Reader: Philosophy and Painting*, Evanston, Ill: Northwestern University Press

Lakoff, G, and Johnson, M (1999) *Philosophy in the Flesh: The Embodied Mind and its Challenge to Western Thought*, New York: Basic Books

McPhee, J (1981) *Basin and Range*, New York: Farrar, Straus and Giroux

McPhee, J (1983) *In Suspect Terrain*, New York: Farrar, Straus and Giroux

McPhee, J (1993) *Assembling California*, New York: Farrar, Straus and Giroux

Mason, B and Berry, LG (1968) *Elements of Mineralogy*, San Francisco: WH Freeman

Merleau-Ponty, M (1962) *The Phenomenology of Perception*, London: Routledge and Kegan Paul

Merleau-Ponty, M (1963) *Phenomenology of Perception*, Smith, C (trans), London: Routledge and Kegan Paul

Merleau-Ponty, M (1968) *The Visible and the Invisible*; followed by working notes edited by C Lefort, translated by A Lingis, Evanston, Ill: Northwestern University Press

Minsky, M (1988) *The Society of Mind*, New York: Simon & Schuster

Raab, T and Frodeman, R (2002) 'What is it like to be a geologist? The phenomenology of geology and its implications' 5(1) *Philosophy and Geography* 69–81

Stoddart, DR (1994) 'Darwin and the seeing eye: iconography and meaning in the Beagle Years' 13 *Earth Science History* 21–34

Vosniadou, S and Ortony, A (1989) *Similarity and Analogical Reasoning*, Cambridge, MA: CUP

Watson, R (1969) 'Explanation and prediction in geology' 77 *Journal of Geology* 488–94

– 12 –

EPILOGUE

HUMANS IN A MINERAL WORLD

Mary Ann Owoc

While assembling the closing remarks for this volume, I found myself returning to my first thoughts on its subject matter. Some years ago now, Don Thieme and I crafted a joint paper for the Society for American Archaeology Meetings entitled 'Post-processual trends and the changing role of geoarchaeology' (Owoc and Thieme 1997). In it, we attempted to highlight the historical, metaphorical nature of earth science descriptions, and urged the discipline of geoarchaeology to augment and revisit its descriptive practices in light of the plurality of conceptions of the mineral world past and present. The somewhat less than enthusiastic response our paper received was unsurprising, not merely in light of the historical context of our remarks, but also because of the epistemological difficulties proposed by our suggestions. The challenges raised in that paper have recently begun to be met by a variety of scholars, and this volume was intended to provide an inspirational forum for work along similar lines. What is at risk in this general endeavour, however, is the unity of the anthropological/archaeological project for making the mineral world meaningful. For example, it might be perceived by some readers of this volume that intellectual advancement is being sought largely through a rejection of past paradigms and methodological achievements, leading inevitably to a fragmented discourse about the mineral world. This would be an incorrect reading of this volume. It might be worthwhile, given our subject matter, to recall the remarks of Brothwell and Higgs (1970: 32) of some 30 years ago regarding the proliferation of separate and independent enquiries arising from restricted practices in environmental archaeology. Describing this as 'a horrid porridge with little meaning', they urged the development of an integrated and critical discipline of archaeological science. We could do worse than follow their suggestions as we seek to further develop our archaeological and anthropological conceptions of the mineral world.

A MATTER OF SCALE?

Stein (1993), in advocating an expanded role for geoarchaeology in archaeological science, described the difficulties of such a project as situated within an issue of scale. Particular to each discipline's encounter with the world, she noted, is a particular scale of interpretation that establishes the boundaries within which inquiry takes place. Both the temporal and spatial elements of such scales become particularly problematic since they influence one's interpretive constructs and methodologies for making the world meaningful and thus constrain what is said in confined avenues. This is an issue of particular relevance since our subject matter – the mineral world – lies at the meeting place of the geosciences and the human sciences – disciplines of inquiry traditionally operating at different spatial and temporal scales.

Interdisciplinary research on the mineral world has characterised the archaeological project almost from its beginnings with the publication of *Geological Evidences of the Antiquity of Man* by Charles Lyell, and the pioneering fieldwork of men like John Frere and Jacques Boucher de Crèvecoeur de Perthes who developed early stratigraphic chronologies for the European Palaeolithic (Gifford and Rapp 1985). Other notable early 19th-century chronological methods based on site stratigraphy and artefact provenience in the US were exhibited by Ephraim Squire and Edwin Davis in their work on mounds throughout the Mississippi Valley. Other early work of this nature that carried on into the 20th century was primarily concerned with large scale chronological issues, particularly those related to glaciation, the formation of various Pleistocene deposits, and human antiquity (Gifford and Rapp 1985; Rapp and Hill 1998; Renfrew 1976).

The most productive merging of archaeology and geoscience has occurred in the last 40 years, facilitated largely by what might be termed a second perceptual expansion in the anthropological/archaeological inquiry towards large scale temporal and spatial questions, coupled with an increased receptivity to the methods and techniques of the natural sciences. This expansion was in part facilitated by the recognition that the archaeological record itself transcends short-term historical occurrences and could therefore inform us about long-term behavioural trends and changes in the environment (Brothwell and Higgs 1970; Butzer 1982). Informed by ecosystems theory and modelling in archaeology (Binford 1962; Clarke 1978; Cornwall 1970: 125; Flannery 1968), the developing geoscientific discourse in archeology has addressed the geomorphological context for human behaviour through techniques designed to investigate natural processes of site formation and preservation, reconstruct the general landscape, and understand the relation of both to prehistoric human behaviour (French 2003: 3–4, Rapp and Hill 1998; Renfrew 1976: 2; Waters 1992: 88). The inherent emphasis upon large scale spatial and temporal processes in such work is understandable in terms of both its disciplinary context and the developmental nature of the geomorphic record itself (Stein 1993: 5).

In contrast, recent years have witnessed a gradual shift in the archaeological/anthropological applications of geoscience techniques towards finer scales of resolution. Stein (1993, 2001), for example, has called for geoarchaeology to focus its interpretative power on the explanation of social principles at the human scale through a critical appreciation of sediments and their definition. Recently, Evans (2003) has outlined how soils should be considered within the social construction of the environment. An attention towards the formation and qualities of anthropogenic soils (Gartner 1992; Rosen 1986; Stein 1985, 1987) has become a part of the mainstream discourse of geoarchaeology, and the recent proliferation of micromorphological analyses has been focused in substantial measure directly upon residential deposits and site lenses (eg Boivin 2000, 2001; French and Lewis 2001; Goldberg and Arpin 2000). Finally, provenance studies increasingly inform questions about the *social* elements of raw material procurement, use and trade (eg Bradley *et al* 1992; Cooney 2002; Hampton 1999), and early pioneering research in ethnominerology (eg Arnold 1971; Ollier *et al* 1971) has been augmented by a wide variety of studies addressing the procurement and definition of various pigments and clays (eg Boivin 2001 and this volume; Gosselain 1998; Livingstone-Smith 2000). Many of these contributions involve, or are informed by, work from these high resolution surveys, but it would be a mistake to assume that the authors are attempting to provide a definitive answer to the question: 'What is the appropriate scale for dealing with the mineral world?' Instead, all have adopted particular perspectives based upon their pragmatic concerns, which involve comprehending how humans appreciate, experience and envision the surface and buried landscape at *many* levels. Seen in this fashion, the issue of scale really becomes subsumed within a larger field related to embodied experience and vision, and our move towards integration begins with a consideration of these factors.

KNOWING THE MINERAL WORLD

As noted above, the context informing much of the interdisciplinary collaboration on the mineral world has been the growth and development of archaeological science. *Science in Archaeology* by Brothwell and Higgs, first published in 1963, defined for a generation of scholars the rationale of a scientific archaeology, its aspirations, and the methodologies for achieving them. In the preface to the revised edition, Grahame Clarke's optimistic vision of scientific development encapsulated the particular ontological world view informing this general field of research. 'Science,' he wrote, 'like the course of civilization itself, advances at an accelerating pace. So also does the application of science to the task of unfolding the unwritten history of man's changing relation to his environment, his advance towards a more complex and rational exploitation of natural resources in the interest of a fuller and more complex social life' (Clarke 1970: 21). Underlying his remarks, and much of the disciplinary practices of the earth science discourse informing archaeology's encounter with the mineral world, is a distinctive earth reasoning that the contributors to this volume have sought to analyse and/or augment. This reasoning may be characterised largely by division, and is informed by a view of separation between humans and the world, where the earth or landscape exists as a backdrop to human action, or as a resource to be exploited (for varied analyses and critiques of this perspective see Ellen and Fukui 1996; Frodeman 2000; Ingold 1996, 2000; Thomas 1993; Tilley 1994). The identification and study of the geomorphic landscape as the non-living component of the *external* environment, the tradition of placing a site in its 'non-cultural' environmental context (which lies at the basis of ecosystem reconstruction), the definition of natural geological entities independent of a prior human presence or naming, and the objective definition and categorisation of sediments and soils are all in one fashion or another informed by this particular view.

By contrast, the papers in this volume take as their starting point a more relational view, in which humans exist not separately from but rather within the mineral world as meaning-giving social organisms, who constitute their landscapes through acts of naming and dwelling, and who in turn are constituted by them in multiple ways that influence their cultural production (Heidegger 1962; Ingold 2000; Tilley 1994). The encounters described in the present volume involve such varied activities as repeated engagements with the materials of the natural landscape, the use of soils and stone in built architecture, the use of pigments and paints for decoration and art, the variable processes of modification involved in material culture creation, and finally the application of technical/specialist knowledge in practices designed to describe the earth. Central to *all* of these practices is an embodied, sensual experience, which is given meaning and significance within a diversity of cultural schemes of perception and sets of social relationships (following Bourdieu 1977: 114–24; Classen 1993; Ingold 2000; and Merleau-Ponty 1962). Appreciating this commonality necessitates an understanding of the process of knowledge production taking place in the many encounters that humans have had with the mineral world over time. Appropriated philosophically (Frodeman 1995, 2000, 2003), the manner in which humans make sense of the earth can be seen to have its basis in practical awareness and lived experience. Frodeman (1995, 2003 and this volume) has described the process of giving meaning and significance to the mineral world as a historically situated *hermeneutic* process, in which subject and object play relational roles. The 'fusion of horizons' (Gadamer 1975: 273) that results in meaningful description or characterisation is a semiotic process, involving a creative, imaginative envisioning of the earth before us (Baker 1999, 2000: 8, Frodeman 2003: 104–10). Through repeated encounters and the acquisition of disciplinary knowledge, the formless and meaningless mineral world acquires shape and significant features over time (Frodeman 2003). In this construction, the contemporary Western geologist is revealed as little different from Hampton's description of an Una tribal quarry expert in Irian Jaya Province, Indonesia who 'understands the rocks just like a healer (medical doctor) knows a patient' (1999: 256). While the bag of disciplinary tools for Una quarry experts may differ in some respects from that of the geologist, geoarchaeologist, or other native expert, the fundamental process in going 'from blindness to insight' (Frodeman, this volume) remains the same, based as it is in knowledge and familiarity. Describing the process of giving meaning to stone and earth as poetic conversation, and the result a production of narrative (Baker 1999; Frodeman 2003 and this volume), properly situates the epistemological claims of all experts within a general human and social process of being-in-the-mineral-world. By so doing, it enlarges the anthropological/archaeological frame of reference to include a multiplicity of significant encounters between people and place from the distant past to the disciplinary present. Within this variety of human engagement with the world past and present lies commonality, and an appreciation of this reveals these multiple engagements as part of a shared human project in the mineral world.

The creative envisioning central to this hermeneutic experience of landscape involves also a practical awareness based in embodied perception (Buttimer and Seaman 1980; Frodeman 2003 and this volume; Tilley 1994; Tuan 1977). Quotidian, seasonal/calendrical and important social and disciplinary activities bring humans into repeated contact with varied soils, sediments, outcrops and stones from their local or regional landscapes. Complementing our well established Western constructs, the papers in this volume indicate the presence of a general, knowledgeable intimacy with the mineral world gained by these encounters, evidenced by common taxonomies involving multiple outcrops, stones and sediments/soils (for other notable examples see Arnold 1971; Boivin 2001; Ellis 1974; Ollier *et al* 1971; Reichel-Dolmatoff 1987). The sorts of practical taxonomies that result from disciplinary awareness and a lived engagement with the earth emerge from the *temporal, physical* nature of these encounters with the mineral world, as well as its own multiple inherent qualities.

Foltz (2000: 30–33), in an attempt to break down barriers between science and other forms of knowledge about the earth, refers to the human encounter with the physical world as 'reconnaissance'. Reconnaissance – orientation in the interest of inhabitation – is basic to all societies since they simultaneously seek to understand and be at home in the world (*ibid*). Central to reconnaissance is, of course, the notion of inhabitation, which itself incorporates a visual intelligence resting upon a 'practical geometry' (Bourdieu 1977: 114–24), and a tactile or kinesthetic type of reasoning, dependent upon the active body in a landscape (Frodeman 2003: 112; Tuan 1977). As mountain, outcrop, beach or stream pebble, exposed sediment, buried soil, or stratigraphic column, the mineral world appeals to us on the basis of its visibility, or hidden nature, its position above and below us, in reference to other materials, and spatially with respect to the paths and places that we visit and experience during the course of our activities. How we both use and meaningfully construct soils and stones, then, is a direct function of the embodied temporal and spatial relationships we have with them (as addressed by Tilley 1994), either as communities engaged in daily routines and ritual practices, or as scientists investigating a site matrix or describing an outcrop, as illustrated in this volume. These relationships involve an awareness of such characteristics as a material's vertical position in relation to the ground surface or horizon (Arnold 1971; Owoc, this volume; Charles *et al,* this volume), or with respect to other elements similarly constructed (Herz and Garrison 1998; Waters 1992; Stein 2001). Further, its elevation or location within culturally defined territories (Brumm, this volume; Ollier *et al* 1971; Taçon, this volume) or geomorphic maps (Herz and Garrison 1998) and in relation to boundary features like the sea or the sky (Scarre 2002; Tilley 1999), also play a role in the development of its classification, meaning and significance. As implied throughout this volume, an ongoing awareness of these basic human perceptions and how they are culturally constructed through a variety of variable human-earth encounters will contribute much to the reflexive development of our descriptive enterprises.

Further, the studies above illustrate the multi-vocal nature of soils and stones by revealing how particular qualities of certain materials are highlighted and meaningfully constructed. The various characteristics of soils and stones addressed in the present work include properties like colour, hardness, reflective properties, texture, structure, the presence or absence of rock fragments or other inclusions, consistency, and plasticity; all of which transgress the boundaries constructed by particular disciplinary or indigenous traditions. Though the acknowledgment and description of these varied physical properties is clearly culturally and temporally variable (as noted by both Boivin and Scarre in this volume), it is nevertheless the case that the materiality of the mineral world appeals to us, as humans interacting with it, on a number of mutual levels. The commonalities inherent in our appreciation of these properties should inspire us to think outside the boundaries of our commonly constructed descriptive frameworks, and consider in what ways the varied properties of these materials influence their cultural appropriation in both Western and other classificatory schemes.

CONCLUSION

The paradigmatic shifts within anthropology, archaeology and the earth sciences over the last 50 years have contributed in a productive way to the development of ever more powerful tools for coming to terms with the human-environment relationship. The diversity of approaches engendered by these shifts, however, has been a mixed blessing, particularly since the sorts of sophisticated understandings of the human project past and present that we currently envision will demand a far more integrated approach. To some extent, of course, conflicting or irreconcilable views of the mineral world may always be an inevitable reality, since the varied nature of cultural and historical circumstances, and the nature of earth itself, make our 'point of maximum grip' (Merleau-Ponty 1962 in Frodeman 2003: 112) a constantly shifting one. Nevertheless, this volume's attempts to more deeply and authentically locate the archaeological and anthropological conversation about the mineral world within a developing dynamic, self aware, human-ecological inquiry have surely deepened and broadened our comprehension of that world in line with the best advances in environmental science.

References

Arnold, DE (1971) 'Ethnominerology of Ticul, Yucatan potters: etics and emics' 36(1) *American Antiquity* 20–40

Baker, VR (1999) 'Geosemiosis' 111 *Geological Society of America Bulletin* 633–46

Baker, VR (2000) 'Conversing with the earth: the geological approach to understanding', in Frodeman, R (ed), *Earth Matters: The Earth Sciences, Philosophy, and the Claims of Community*, Upper Saddle River: Prentis Hall

Binford, L (1962) 'Archaeology as anthropology' 28 *American Antiquity* 217–25

Boivin, N (2000) 'Life rhythms and floor sequences: excavating time in rural Rajasthan and Neolithic Çatalhöyük' 31(3) *World Archaeology* 367–88

Boivin, N (2001) *Archaeological Science as Anthropology: Time, Space and Materiality in Rural India and the Ancient Past*, unpublished PhD thesis, Department of Archaeology, University of Cambridge

Bourdieu, P (1977) *Outline of a Theory of Practice*, Nice, R (trans), Cambridge: CUP

Bradley, R, Meredith, P, Smith, J and Edmonds, M (1992) 'Rock physics and the Neolithic axe trade in Great Britain' 34(2) *Archaeometry* 223–33

Brothwell, D and Higgs, E (1970) *Science in Archaeology: A Survey of Progress and Research*, New York: Praeger

Buttimer, A and Seaman, D (eds) (1980) *The Human Experience of Space and Place*, New York: Palgrave Macmillan

Butzer, K (1982) *Archaeology as Human Ecology: Method and Theory for a Contextual Archaeology*, Cambridge: CUP

Clarke, D (1978) *Analytical Archaeology*, 2nd edn, London: Methuen

Clarke, G (1970) 'Foreword', in Brothwell, D and Higgs, E (eds), *Science in Archaeology: A Survey of Progress and Research*, New York: Praeger

Classen, C (1993) *Worlds of Sense: Exploring the Senses in History and Across Cultures*, London: Routledge

Cooney, G (2002) 'So many shades of rock: colour symbolism and Irish stone axeheads', in Jones, A and McGregor, G (eds), *Colouring the Past: The Significance of Colour in Archaeological Research*, London: Berg

Cornwall, IW (1970) 'Soil, stratification, and environment', in Brothwell, D and Higgs, E (eds), *Science in Archaeology: A Survey of Progress and Research*, New York: Praeger

Ellen, R and Fukui, K (eds) (1996) *Redefining Nature: Ecology, Culture and Domestication*, Oxford: Berg

Ellis, FH (1974) 'The Hopi: their history and use of lands', in Horr, DA (ed), *Hopi Indians*, New York: Garland

Evans, J (2003) *Environmental Archaeology and the Social Order*, London: Routledge

Flannery, K (1968) 'Archaeological systems theory and early Mesoamerica', in Meggers, BJ (ed), *Anthropological Archaeology in the Americas*, Washington, DC: Anthropological Society of Washington

Foltz, BV (2000) 'Inhabitation and orientation: science beyond disenchantment', in Frodeman, R (ed), *Earth Matters: The Earth Sciences, Philosophy, and the Claims of Community*, Upper Saddle River: Prentis Hall

French, CAI (2003) *Geoarchaeology in Action: Studies in Micromorphology and Landscape Evolution*, London: Routledge

French, CAI and Lewis, H (2001) 'The Fengate Depot site: the micromorphological analysis', in Pryor, R (ed), *The Flag Fen Basin: Archaeology and Environment in a Fenland Landscape*, Archaeological Reports, London: English Heritage

Frodeman, R (1995) 'Geological reasoning: geology as an interpretive and historical science' 107(8) *Geological Society of America Bulletin* 960–68

Frodeman, R (ed) (2000) *Earth Matters: The Earth Sciences, Philosophy, and the Claims of Community*, Upper Saddle River: Prentis Hall

Frodeman, R (2003) *Geo-logic: Breaking Ground Between Philosophy and the Earth Sciences*, Albany: State University of New York Press

Gadamer, H-G (1975) *Truth and Method*, London: Sheen and Ward

Gartner, WG (1992) 'Soils and sediments as artifacts in prehistoric Wisconsin', in Foss, JE, Timson, ME and Morris, MW (eds), *Proceedings of the First International Conference on Pedo-Archaeology*, Knoxville: University of Tennessee Press

Gifford, JA and Rapp, G, Jr (1985) 'History, philosophy and perspectives', in Gifford, JA and Rapp, G, Jr (eds), *Archaeological Geology*, New Haven: Yale UP

Goldberg, P and Arpin, T (2000) 'Micromorphological analysis of sediments from Meadowcroft Rockshelter, Pennsylvania: implications for radiocarbon dating', *Journal of Field Archaeology* 325–42

Gosselain, OP (1998) 'Social and technical identity in a clay crystal ball', in Stark, MT (ed), *The Archaeology of Social Boundaries*, Washington, DC: Smithsonian University Press

Hampton, OW (1999) *Culture of Stone: Sacred and Profane Uses of Stone Amongst the Dani*, College Station: Texas A and M University Press

Heidegger, M (1962) *Being and Time*, Macquarrie, J and Robinson, E (trans), New York: Harper and Row

Herz, N and Garrison, EG (1998) *Geological Methods for Archeology*, New York: OUP

Ingold, T (1996) 'Hunting and gathering as ways of perceiving the environment', in Ellen, R and Fukui, K (eds), *Redefining Nature: Ecology, Culture and Domestication*, Oxford: Berg

Ingold, T (2000) *The Perception of the Environment*, London: Routledge

Livingstone-Smith, A (2000) 'Processing clay for pottery in Northern Cameroon' 42(1) *Archaeometry* 21–42

Merleau-Ponty, M (1962) *Phenomenology of Perception*, London: Routledge

Ollier, CD, Drover, DP and Godlier, M (1971) 'Soil knowledge amongst the Baruya of Womenara, New Guinea' 42(1) *Oceania* 33–41

Owoc, MA and Thieme, D (1997) 'Post-processual trends and the changing nature of geoarchaeology', Paper presented at the 62nd Annual Meeting of the Society for American Archaeology, Nashville, Tennessee

Rapp, G and Hill, CL (1998) *Geoarchaeology: The Earth Science Approach to Archaeological Interpretation*, New Haven: Yale UP

Reichel-Dolmatoff, G (1987) 'The great mother and the Kogi universe' 13(1) *Journal of Latin American Lore* 73–113

Renfrew, C (1976) 'Archaeology and the earth sciences', in Davidson, DA and Shackley, ML (eds), *Geoarchaeology*, London: Duckworth

Rosen, AM (1986) *Cities of Clay: The Geoarchaeology of Tells*, Chicago: University of Chicago Press

Scarre, C (2002) 'Epilogue: colour and materiality in prehistoric society', in Jones, A and McGregor, G (eds), *Colouring the Past: The Significance of Colour in Archaeological Research*, Oxford: Berg

Stein, JK (1985) 'Interpreting sediments in cultural settings', in Stein, JK and Farrand, WR (eds), *Archaeological Sediments in Context*, Orno: Center for the Study of Early Man

Stein, JK (1987) 'Deposits for archaeologists', in Schiffer, M (ed), *Advances in Archaeological Method and Theory Vol 11*, New York: Academic

Stein, JK (1993) 'Scale in archaeology, geosciences, and geoarchaeology', in Stein, JK and Linse, A (eds), *Effects of Scale on Archaeological and Geoscientific Perspectives*, GSA Special Paper 283

Stein, JK (2001) 'Archaeological sediments in cultural environments', in Stein, JK and Farrand, W (eds), *Sediments in Archaeological Context*, Salt Lake City: University of Utah Press

Stein, JK and Farrand, W (2001) *Sediments in Archaeological Context*, Salt Lake City: University of Utah Press

Thomas, J (1993) 'The politics of vision and the archaeologies of landscape', in Bender, B (ed), *Landscape, Politics and Perspectives*, Oxford: Berg

Tilley, C (1994) *A Phenomenology of Landscape: Places, Paths and Monuments*, Oxford: Berg

Tilley, C (1999) *Metaphor and Material Culture*, Oxford: Blackwell

Tuan, Y-F (1977) *Space and Place: The Perspective of Experience*, Minneapolis: University of Minnesota Press

Waters, MR (1992) *Principles of Geoarchaeology: A North American Perspective*, Tucson: University of Arizona Press

INDEX